SPACE TELESCOPE SCIENCE INSTITUTE

SYMPOSIUM SERIES : 1
Series Editor S. Michael Fall, Space Telescope Science Institute

Stellar Populations

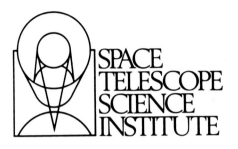

SPACE
TELESCOPE
SCIENCE
INSTITUTE

Stellar Populations

Proceedings of the Stellar Populations Meeting
Baltimore, 1986 May 20 - 22

Edited by

COLIN A. NORMAN
Space Telescope Science Institute, Baltimore

ALVIO RENZINI
Osservatorio Astronomico, Bologna

MONICA TOSI
Space Telescope Science Institute, Baltimore
Osservatorio Astronomico, Bologna

Published for the
Space Telescope Science Institute

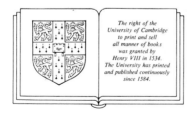

The right of the
University of Cambridge
to print and sell
all manner of books
was granted by
Henry VIII in 1534.
The University has printed
and published continuously
since 1584.

CAMBRIDGE UNIVERSITY PRESS

Cambridge

London New York New Rochelle

Melbourne Sydney

Published by the Press Syndicate of the University of Cambridge
The Pitt Building, Trumpington Street, Cambridge CB2 1RP
32 East 57th Street, New York, NY 10022, USA
10 Stamford Road, Oakleigh, Melbourne 3166, Australia

Printed in Great Britain at the University Press, Cambridge

British Library cataloguing in publication data

Stellar Populations Meeting *(1986 : Baltimore)*
Stellar populations : proceedings of the Stellar Populations Meeting, Baltimore, 20-22
May 1986. - (Symposium Series/Space Telescope Science Institute; 1)
1. Stars - Populations
I. Title II. Norman, C.A. III. Renzini, A. IV. Tosi, M. V. Space Telescope Science
Institute VI. Series 523.8 QB801

Library of Congress cataloging in publication data available

ISBN 0 521 33380 6

CONTENTS

FOREWORD

The idea of a series of annual topical seminars to be held at Space Telescope Science Institute arose in discussions with Lennox Cowie and other members of the scientific staff about two years ago. It seemed clear that after the launch of the Hubble Space Telescope (now estimated for late 1988), there would need to be a forum for the presentation and discussion of the new findings of the Hubble Space Telescope. Prior to the launch these seminars would provide a high level overview of the status of knowledge in particular fields where observations with this telescope could be expected to provide significant advances.

The format that we chose was that of a rather informal workshop, with invited reviews, poster papers, and particular emphasis on providing ample time for discussion and interaction. The first symposium, "Deep Observations of the Formation and Evolution of Galaxies" was held on May 1-2, 1985, and Richard Kron, while Visiting Scientist at the Institute, played a major role in its organization. The meeting was extremely well attended and quite useful to specialists in the field, and it became obvious that publication of the results would have proven useful to astronomers generally. Consequently an agreement was contracted between Cambridge University Press and the Space Telescope Science Institute for the rapid publication of the material of future Symposia.

It gives me great pleasure to introduce this volume on "Stellar Populations" as the first publication in the "Space Telescope Science Institute Symposium Series", which will be forthcoming in the next several years. Monica Tosi, Visiting Scientist at the Institute, has played a major role in its organization. Sponsorship of these Symposia is carried out in part as a fulfillment of our responsibilities under NASA Contract NAS5-26555, with additional funding provided through the ST ScI Associates Program

Riccardo Giacconi

PREFACE

The understanding of fundamental processes such as the formation, internal organization and evolution of galaxies ultimately relies on the study of the product of these complex activities, namely the stellar content of galaxies. More specifically, it is founded on the determination of stellar metallicities, ages, positional and dynamical properties, which indeed make up our current conception of "Stellar Populations". It is certainly this perspective of understanding the origins of stellar systems which gives scientific momentum and motivation to a number of otherwise disparate studies in this field, with their apparently endless details and intricacies.

From a technical point of view, the study of stellar populations naturally flows in two streams, on the one hand star by star investigations of "resolvable" stellar populations, and on the other integrated light studies, even where the first is not possible any more. The former endeavour has recently greatly expanded, thanks to the advent of linear detectors and sophisticated data processing. Further major advances in this sector are eagerly expected from the Hubble Space Telescope. The integrated approach, for much the same reasons, has brought the frontier of observed galaxies to very high redshifts, very close to the formation epoch. Again, the Hubble Space Telescope will give sharper contours to the fuzzy young galaxies of today.

It was with the perception of the rapid growth of this field in recent years, and of the widespread expectations from the Hubble Space Telescope, that the Space Telescope Science Institute decided to organize this Symposium on Stellar Populations.

More than one hundred scientists from all over the world attended the Meeting. We are grateful to all of them for having contributed so much to its final success by presenting their new results in posters or short communications (collected in a special issue of the Space Telescope Science Institute Preprint Series) and by generating the lively and stimulating discussion that followed each talk.

We are deeply indebted to Barbara Eller for her excellent arrangement of all the accommodations and social events during the Meeting, and for her help in the overall organization of both the Meeting and the Proceedings. We thank Janet Boss, Carmella Quattro, Karen Vrana, and Loretta Willers for always being ready to help, and we specially appreciated the skill of Marilyn Bell in transcribing the discussions from tapes, and of Dorothy Schlogel in preparing camera ready materials.

We dedicate this book to the memory of Beatrice Tinsley whose remarkable intelligence and personality profoundly influenced this entire subject.

Colin A. Norman

Alvio Renzini

Monica Tosi

Beatrice Tinsley, Yale University
Yale University Office of Public Information

BEATRICE MURIEL TINSLEY--A DEDICATION AND APPRECIATION

James E. Gunn
Princeton University, Princeton, NJ 08544

Nine years ago this month there was held at Yale a meeting about galaxies that is universally regarded as a kind of watershed, a meeting which signalled to the more optimistic of us a coming of age of a subject which had been largely taxonomic, and to all of us a realization of the immense complexity of the field.

That this meeting, at which we will largely be addressing the questions framed so well at that earlier one, be dedicated to the memory of the person who made the Yale meeting happen and did so very much to make the revolution in extragalactic astronomy happen, is happy and fitting.

It is a tribute to the vigor of the field that many of the participants at this Baltimore meeting are too young to have had the opportunity to know Beatrice or work with her, though her work, which permeates the subject of populations in galaxies, is well known to us all. She is responsible for the first serious attempts at the synthesis of the stellar populations of galaxies and their evolution, the invention of the widely used technique of evolutionary synthesis, which indeed made early attempts at population synthesis possible, countless contributions to the understanding of chemical evolution and population evolution in general, and very early recognition that environmental interactions can have drastic effects on the structure and evolution of galaxies. Her techniques and nomenclature are used by us all, and so prodigious and influential was her work that she will certainly be remembered for her scientific contributions for a very long time.

Four years after the Yale conference she was dead, the victim of a melanoma she already knew she had when the conference occurred. We and astronomy are very much the poorer for the loss of her, and I have no doubt that more answers would be apparent here for the questions asked at that conference had she lived. She was a truly remarkable person, with a zest for her subject I have never seen before or since. Collaborating with her, which I was privileged to do on many occasions, was more than a bit like grappling with a whirlwind-- nasty questions and problems were not always answered, but they were, to be sure, exposed, and if they were left unanswered, it was never for lack of invention and effort. She was indefatiguable to the end; she made a legend in her last days to which we could all usefully but probably hopelessly aspire. Obviously terminally ill, weakened by radiation and chemotherapy, she set up her final office about a year before her death in the Yale infirmary, an institution more used to dealing with flu, broken limbs, and allergies than terminal cancer patients. But she was far more interested in getting on with her work and finishing the many papers (and starting a few) that were in the works than worrying about dying, and a cancer ward seemed to her not the ideal place to do that. The staff, at first worried about the effect of housing such a seriously ill patient on others, soon found that she was the liveliest of the lot, often taking it upon herself to cheer others infinitely healthier than she. Most of every day her room was filled with astronomers from all over the world, some, to be sure, well-wishers, but most at work writing papers with her. In the last months the cancer invaded her brain and began to impair her motor functions but not her intellect. She lost the use of her right arm, so she learned to write with her left, so she could go on, and go on she did, to the very end.

All of us who knew her miss her terribly, and I think the field will never be quite the same. I am very happy that the organizers of this meeting decided to dedicate it to her, and grateful for the opportunity to say these few words and make a little introduction to those of you who did not know her.

INTRODUCTION AND OVERVIEW

Leonard Searle
Mount Wilson and Las Campanas Observatories,
Carnegie Institution of Washington
Pasadena, California

ARGUMENT

The eyes are more exact witnesses than the ears.

Heracleitus of Ephesus
(Diels, Fragmente B101a)

The Present Situation

The fundamental fact that underlies the concept of stellar populations is that the stellar content of different galaxies, or of subsystems
within a galaxy, can be very different. Baade's sharp dichotomy was
between regions where star-formation is active and those where it has
long since ceased. For a clear summary of his views, see Baade (1958).
We now recognize his Population II - the globular clusters, the Galactic
bulge, the resolved disk of the Andromeda spiral, and the dwarf spheroidals - as a miscellany of heterogeneous populations. We know, too, that
Population I is both chemically and kinematically heterogeneous.

What would an idealized population description of a stellar system
look like? It would involve, I think, a subdivision of the stellar system into significant subsystems, defined by their kinematic properties,
or by their spatial distribution, and it would require the determination
of the age distribution and the abundance distribution of each subsystem.

This idealized population description is useful to keep before us
because it contrasts so sharply with what we actually know at present
regarding any recognized stellar population. It also points out something else--namely, that few of us would be interested in the subject at
all if such a description were the aim of the study of stellar populations.
What drives the subject is an interest in origins, in the processes that
lead to things being the way that they are.

What is the present state of the study of Stellar Populations? On
the descriptive side, much progress is being made and there are exciting
new results on the abundance distributions and star-formation histories
of the nuclear bulge of the Galaxy, of the globular cluster systems and
spheroid stars of our Galaxy and other Local Group spirals, and of the
dwarf spheroidal galaxies. The first results are becoming available
concerning stellar populations in the galaxies belonging to remote
clusters. We confidently anticipate the wealth of new observational fact
that will come when the Hubble Space Telescope is applied to these
problems.

There is slower, but nonetheless real progress in understanding the processes that give rise to stellar populations. But the new observational results serve to emphasize how inadequate that understanding is. What has been newly discovered has been, in every case, I believe, completely unanticipated, and this points up, very emphatically, that our present ideas about how galaxies form are without predictive power.

Understandings

The subject of Stellar Populations comes to us with a history that brings with it, not only a rich heritage of facts and concepts, but also, no doubt, some misconceptions and unresolved contradictions, in which present controversies have their roots.

The fundamental kinematic dichotomy between disk and halo of the Galaxy, recognized by Lundmark, was given   its interpretation in terms of Galactic rotation by Lindblad (1925), in a paper that introduced a model of the Galaxy as a superposition of progressively flattened and progressively more rapidly rotating subsystems. The model involved multiple subsystems because Lindblad was concerned to explain the kinematic properties of intermediate populations. He cited planetary nebulae as being objects with kinematics and space distribution that distinguish them from typical stars of the Milky Way on the one hand and from globular clusters on the other.

This successful kinematic model was given a hypothetical evolutionary interpretation in the early 1950's, an interpretation briefly summarized, for example, in von Weizsacker (1955). This interpretation, persuasive and influential, served to organize all thought on Stellar Populations, for the better part of a quarter of a century. Any account of progress in understanding the origin of stellar populations must start from the synthesis, based on these ideas, developed at the 1957 Conferenc on Stellar Populations, sponsored by the Vatican Observatory.

It seems to me to be important to recall the context in which these ideas arose, and to appreciate the received opinion of that time. Giving an account of galaxy formation in his article on cosmogony, written in the early 1960's for the Encyclopedia Britannica, Gamow wrote "It can be shown that, as long as they remained gaseous, the newly formed protogalaxies must have been undergoing steady contraction . . .. During these early contractive stages, protogalaxies must have assumed their character istic elliptical shapes, since, as was shown by Jeans, various observed galactic forms correspond to equilibrium configurations of rotating gas masses, and since, on the other hand, after the formation of stars the galaxies must have lost the internal coherence of material necessary for assuming regular geometrical shapes." and he went on to emphasize that "the regular shapes of the galaxies as seen at present cannot be ascribed to the statistical interaction between individual stars."

My purpose in citing these words is, of course, to emphasize the revolution that has taken place in the last decade concerning probable

modes of the formation of galaxies. The studies of the rotation of
elliptical galaxies by Illingworth and by Binney, among others, furnish
the strongest evidence we have that cosmogony is a science, and that at
least some of its assertions can be refuted. The recognition that the
observed forms of galaxies do not, in general, derive from the forms of
pre-existent gaseous masses is one of three important developments in
cosmogony that have taken place since Gamow wrote the words that I have
quoted.

The recognition that the motions of stars do not, in general, derive
from the motions of the gas masses out of which they formed, is the
second of these developments. Numerical experiments have familiarized
us with the unexpected consequences of collective dynamical processes,
and we no longer suppose two-body interactions to be the only engine
that drives the evolution of the kinematic properties of stellar systems.
Whatever the detailed mechanisms involved, the observed correlations between age, velocity dispersion and scale height of stars in the solar
neighborhood put beyond doubt that the secular acceleration of stars
plays a major role in causing the phenomena that Lindblad's model so
well accounts for. The basic kinematic properties of stellar disks,
such as the radial gradient in velocity dispersion, also surely have
their origin in kinematic evolution, rather than in the motion of the
gas clouds out of which the disk formed.

Stressing the importance of kinematic evolution, and writing in
1972, Freeman (1975) summarized what remained of the old view, after
this importance is recognized: "For the purposes of our formation picture, it is then probably meaningful to distinguish only two independent
populations, defined by the dynamical processes which gave to each its
random energy: the disk (including stars of all ages) and the halo. In
other words, while the Galactic subsystems with varying degrees of flattening (as recognized by Lindblad) are obviously dynamically significant,
it may be that only two classes of these subsystems (i.e., the disk and
the halo) are cosmogonically significant."

The third important development in cosmogony has been the recognition that the lives of galaxies cannot, in general, be assumed to be
uneventful, nor to be lived in isolation. For rather more than a
decade, Alar Toomre has been persuasively insisting that galaxy interactions are frequent now, must have been more so in the past, and must
have played a decisive role in determining the structure of at least
some galaxies.

These three developments rest on solid bodies of both observational
and theoretical work and they have severely undermined the persuasive-
ness of the old synthesis. That synthesis, accordingly, no longer
commands the general assent that it once did. There is now no generally
accepted view of how galaxies form, nor, in consequence, of how stellar
populations arise. Despite this, I think that we might agree about some
of the processes involved. It seems sensible (at least, to someone of
a uniformitarian cast of mind) to attempt to account for the population
properties of galaxies in terms of a minimal list of processes for which

there is some convincing observational evidence. My occamized list of contributing processes would go something like this:

(1) The collapse of rotating gaseous masses: which is needed to account for the flattened disks of spirals.

(2) Star-formation: which is something observed to occur only in collapsed gaseous disks, and which proceeds particularly vigorously when these disks collide.

(3) The processes causing the kinematic evolution of disk stars, including the energetic ones needed to account for the existence of young high-velocity stars in the halo.

(4) The interaction and merging of galaxies: with the associated kinematic evolution, enhanced star-formation and removal of gas.

All of these processes undeniably play some part in producing the population structure of at least some galaxies, and it is unclear that anything else is needed. Pitifully little is known about any of them. Still less is known about their relative importance. Clearly, the study of stellar populations has only just begun.

Three Old Problems

I shall conclude by briefly mentioning three old and thorny problems that I hope will be illumined at this conference. The first is the nature of the intermediate populations in the Galaxy. The second is the question of the uniformity of the initial mass function. The third is the problem of interpreting the integrated spectra of stellar systems.

First then, consider Freeman's conclusion (quoted earlier), that there are two cosmogonically significant classes of subsystem in the galaxy: those that got their random energies by acceleration after they were formed in the disk, and those that got their higher random energies in some other way. One question I want to raise is whether there is any observational evidence that supports this very plausible dichotomy. The other question is related: how does one go about assigning an observed intermediate population, for example the metal-rich RR Lyrae variables, to one or other of Freeman's classes? To put the matter yet another way: do intermediate populations exist in the Galaxy that we can be quite sure did not originate in the disk, and that did not reach their present high energies by subsequent secular acceleration?

The problem of the history of the initial mass function has long seemed to be intractable. If we infer anything from observation regarding the history of star-formation in a galaxy, or if we infer anything from an abundance distribution regarding galactic nucleosynthesis, we cannot avoid making strong assumptions about the history of the initial-mass-function. The common assumption of uniformity is nothing but a methodological convenience, although it is arguably better to make that assumption than to speculate at random. Are there any glimmerings from observation or theory that can guide us here?

Finally, integrated light. If the subject of stellar populations is to be extended beyond the Local Group, the age distribution of the stars that make up a galaxy, and the abundance distribution of these same stars will need to be derived from the analysis of integrated light. If I am not mistaken, this subject has a bad reputation. Too much has been claimed, and too few have been persuaded. There is a need to improve the credibility of such analyses, and, happily, there is a clear way forward. The simplest problem in this subject is the determination of the age and abundance of an individual star cluster from a study of its integrated light. If that cannot be done, what else can be believed? The star clusters of the Magellanic Clouds have ages and abundances that can be derived from color-magnitude arrays and from the spectra of individual stars. These clusters provide the analysts of integrated light with a rare opportunity to make predictions that can be verified or refuted. Why is this opportunity neglected?

References

Baade, W. (1958). Spec. Vat. Ric. Ast. 5, 3.
Freeman, K. (1975). In Galaxies and The Universe, ed. A. Sandage,
    M. Sandage, and J. C. Kristian, p. 496, Chicago: University of
    Chicago Press.
Lindblad, B. (1925). Ark f. Mat. Ast. Fys. 19A, No. 21.
von Weizsacker, C. F. (1955). Zs. f. Ap. 35, 252.

THE BULGES OF THE GALAXY AND M31

Jeremy Mould
Palomar Observatory
Caltech 105-24, Pasadena CA 91125
U.S.A.

ABSTRACT The bulges of these two galaxies may differ from their globular cluster systems in a number of important ways. This review highlights the evidence that the mean metallicities of the bulges are higher, and that the bulges harbor, possibly as a separate component, a population traced out by luminous asymptotic giant branch stars and OH/IR sources.

1. INTRODUCTION

When we talk about the "bulge" of a galaxy, what we have in mind is the spheroidal component, whose luminosity changes relative to the disk continuously along the Hubble sequence. If we ask ourselves what we want to learn about the bulges of galaxies, observers will answer: the density law, the chemical composition gradient, and the kinematics. But what we really want to know, of course, is how bulges formed. Did they form dissipatively as one generation of star formation followed another, while the protogalactic gas settled into equilibrium ? The trademark of such a process is a chemical gradient. Or did the stars form first, well out of gravitational equilibrium, and then undergo violent relaxation ? Or did the bulge grow gradually by accretion of gas clouds, which were drawn into the potential well and formed stars ? Chemical gradients are not really a necessary consequence of the latter two processes.

Surprisingly, although Population II was identified in the Galaxy long ago, rather little is known about the bulge of the Milky Way as a whole. If we take the Bahcall, Schmidt and Soneira (1983 – BSS) model of the Galaxy as a departure point, we find the bulge represented as a central component of $1 \times 10^{10} M_\odot$ concentrated inside 1 kpc, plus a spheroid of $2.5 \times 10^9 M_\odot$ within 60 kpc. By comparison with these components the globular cluster system of the Galaxy is rather small beer. Yet it has received overwhelmingly more attention than the other components. It is desirable in future work to right the balance, and this will be the theme of the discussion presented here.

2. ONE OR TWO COMPONENTS ?

If we take the BSS model, add together the gravitational potential of the central and spheroidal components, and take the radial derivative, we obtain the density distribution shown in Figure 1. (One modification to the model was actually made: the galactic rotation curve was taken to be flat inside 1 kpc, eliminating the ambiguous central bump in the rotation curve.) This density distribution is tolerably well fit by an $r^{-3.5}$ power law. Such a power law also fits the density distribution of globular clusters (Zinn 1985). So the question arises: does one really need a central component in the Galaxy to balance the rotation curve ?

The following are some non-dynamical arguments in favor of the reality of the central component:

1. The bulge of the Galaxy is very well delineated in the IRAS 12μ source counts. These sources are believed to be OH/IR stars for the most part. Habing *et al* (1986) fit a de Vaucouleurs law to the radial distribution and find an effective radius which corresponds to $r_e = 0.7$ kpc, dramatically different from the classical spheroid's $r_e = 3$ kpc (de Vaucouleurs and Pence 1977).

2. Blanco and Blanco (1986) have shown that the surface density of late M stars in the optical windows on the nuclear bulge falls off with radius much more rapidly than expected from standard surface brightness laws.

3. The density distribution of RR Lyrae stars is shallower than $r^{-3.5}$ within the solar circle according to Saha (1985), The Lick RR Lyrae Survey has now detected over 200 variables located from 4 to 35 kpc galactocentric radius (Kinman 1986).

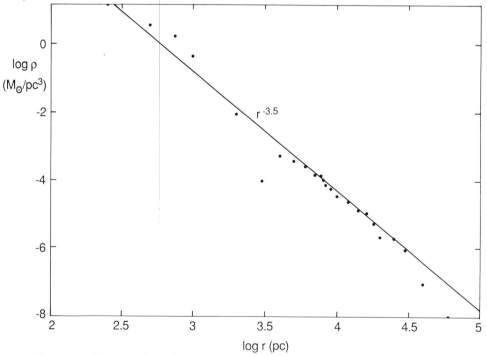

Figure 1. The combined density distribution of the spheroid and the central component in the BSS model of the Galaxy. The central component was modified (see text). Numerical differentiation is the source of the unexpected scatter in the points.

4. The projection of the density distribution of Figure 1 into a surface density profile is shown in Figure 2. It is apparent that the central kpc is poorly fitted by the $r_e = 3$ kpc de Vaucouleurs law. Note, however, that a higher value of M/L in the center of the Galaxy would circumvent this problem. Larson (1986 and this volume) has suggested that preferential formation of massive stars in the center of the Galaxy would provide for this and solve a number of other problems.

5. Conceivably, one could find a theoretical basis for a two component model of the Galactic bulge in a dissipative theory for the central component and an accretion theory for the outer halo.

Some arguments to the contrary are as follows:

6. By raising the local spheroid density considerably above the Schmidt (1975) Pop.II/Pop.I ratio, Caldwell and Ostriker (1980) are able to fit the Galactic rotation curve with a Hubble law and constant M/L (see also Binney 1985). Eggen (1983) provides evidence for such a higher local ratio, which is disputed by Bahcall and Casertano (1986).

7. The tracers of the density of the spheroid discussed in points 1 to 3 above are biased. Late M giants, which are likely progenitors of OH/IR stars, favor regions of high metallicity; RR Lyrae stars favor regions of low metallicity. A metallicity gradient in the nuclear bulge might produce the observed M giant concentration.

8. These different tracers do not seem to have radically different kinematics.

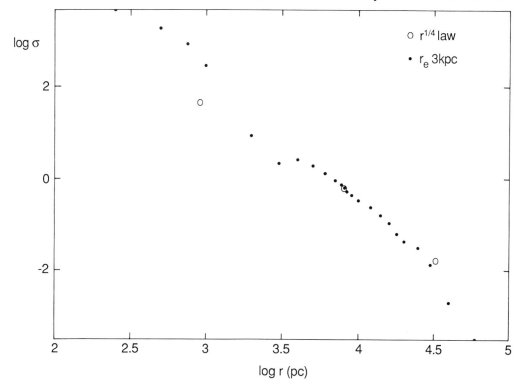

Figure 2. The surface density distribution corresponding to Figure 1, (solid symbols). The open circles show a de Vaucouleurs law with effective radius 3 kpc fitted locally.

Whichever of these points of view prevails, it seems probable that the standard Bahcall and Soneira (1980 – BS) model without a central component underestimates the spheroid density in the central parts of the Galaxy. Star counts in the windows on the nuclear bulge are clearly the way to settle that particular issue.

3. KINEMATICS AND METALLICITY RANGE OF THE NUCLEAR BULGE

The premises of this project initiated by Whitford and Rich (1982) are that K giants form an unbiased sample over the range of metallicities present in the bulge, and that most of the K giants in Baade's window are bulge stars. In the BS model bulge giants and disk giants are approximately equally common at that latitude. Inclusion of the BSS central component increases the ratio to more than 6:1.

Recent results are reported at this meeting by Rich (see Figure 3), and can be summarized as follows. The metallicity distribution in Baade's window is quite different from that of Galactic globular clusters and subdwarfs in the solar neighborhood (cf. Hartwick 1984). The simple model of chemical enrichment with $2Z_\odot$ yield is a good fit to the data. When the sample is divided into 3 subsets by abundance, the subset of 21 stars with $[Fe/H] > 0.3$ has $\sigma = 92 \pm 14$ km/sec and the metal poor subset of 16 stars with $[Fe/H] < -0.3$ has $\sigma = 126 \pm 22$ km/sec. An intermediate set of 16 stars has $\sigma = 97 \pm 17$ km/sec. Kinematic differences between metal rich and metal poor subsamples in Baade's window are detectable at only the 1σ level in the present data. The slightly lower velocity dispersion of the most metal rich stars could reflect (a) a steeper density distribution, (b) a higher level

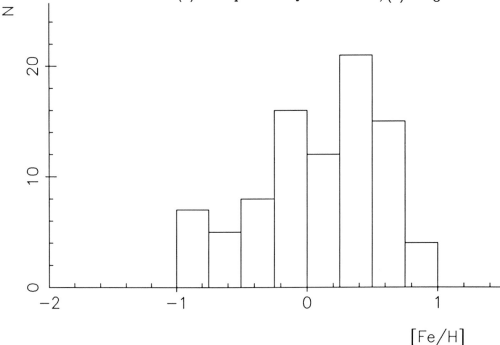

Figure 3. The distribution of metallicity among 88 K giants in Baade's window (Rich 1986).

of rotational support, or (c) small number statistics.

Clearly, a number of interesting issues are emerging from this work. To follow up on these, future studies should include: determination of the density laws for metal rich and metal poor stars, including the disk "field" stars, measurement of their v_{rot}/σ, and analysis of larger samples to achieve full statistical significance.

4. THE BULGE OF M31

The high inclination of the disk of M31 to the plane of the sky makes it possible to investigate both the gradient in mean metallicity and the metallicity dispersion profile of M31's bulge from the properties of the giant branch. A study of one field 7 kpc out on the minor axis has been published by Mould and Kristian (1986). Figure 4 is a color magnitude diagram of a second field obtained this time with 4-SHOOTER (Gunn et al 1982) at the 5-m telescope. Photometry, obtained using similar techniques, is on the Gunn system, fully described by Schneider, Hoessel and Gunn (1983). The field is located 12 kpc out on the minor axis, and includes the cluster G323 of Sargent et al (1977). In this diagram we see the first 3 to 4 magnitudes of the giant branch, stopping short of the horizontal branch. Better seeing conditions have enabled van den Bergh and Pritchet (1986) to detect RR Lyrae stars in a field close to this one. On this figure are superposed giant branches for M92 and M71, observed in the same run with 4-SHOOTER, and shifted to

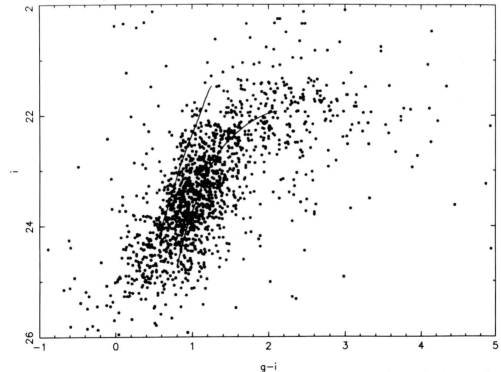

Figure 4. CM diagram of a field in M31 12 kpc out on the SE minor axis. Also shown: giant branches of M92 (left) and M71 (right).

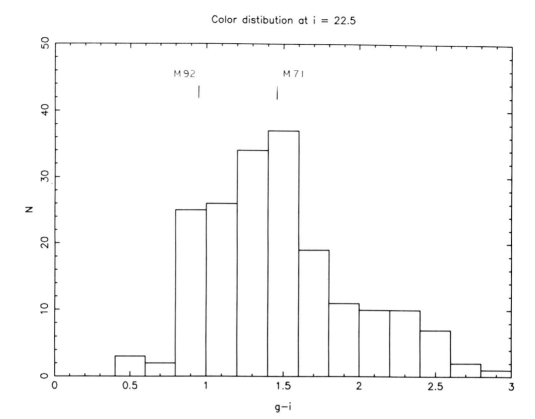

Figure 5. The color distribution on the giant branch of M31 12 kpc on the SE minor axis.

$(m-M)_0 = 24.4$ and $E(B-V) = 0.08$. We can learn about the metallicity distribution in this field by plotting the color histogram at $i = 22.5 \pm 0.25$ (Figure 5). The colors at which the shifted giant branches of the Galactic globulars cross $i = 22.5$ indicate that this field is characterized by a rather metal rich population. Average observational errors at this magnitude are thought to be less than a bin in Figure 5, although detailed testing with artificial stars remains to be done.

Stepping closer in on the SE minor axis, we see another field (including cluster G263) at 5 kpc projected galactocentric distance (Figure 6a). This is a fairly crowded field, and the photometry was carried out using the DAOPHOT point-spread-function fitting program (Stetson 1985). Figure 6b is the corresponding color histogram. Finally, at 20 kpc on the same minor axis there is Figure 7, in which the giant branch can barely be made out for field stars, and its color histogram, Figure 8.

Although it is too early to discuss the metallicity dispersions seen here, it is

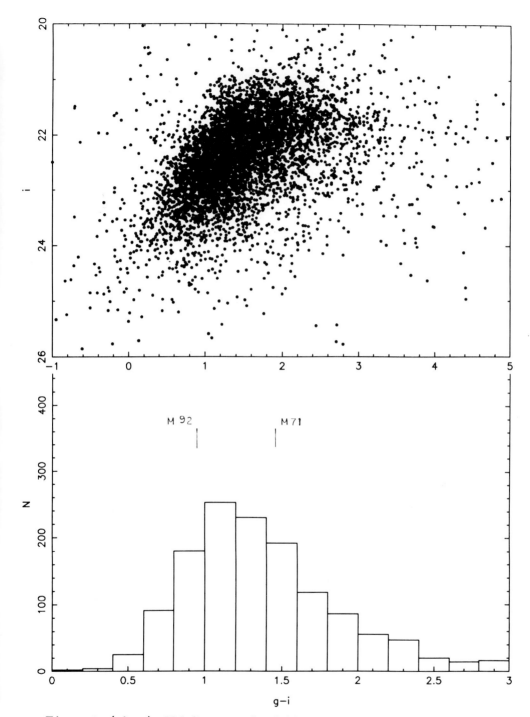

Figure 6a (*above*). CM diagram of a field 5 kpc out on the SE minor axis of M31 and *b.* (*below*) the color distribution at $i = 22.5$ on this giant branch.

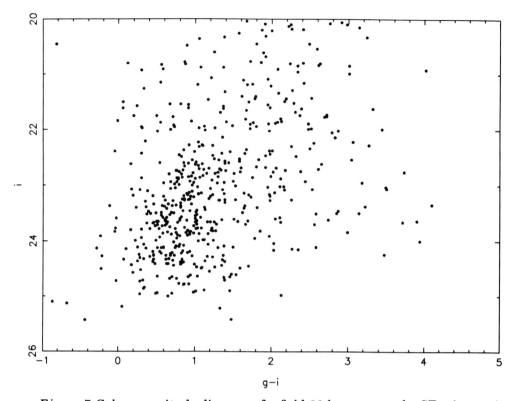

Figure 7 Color magnitude diagram of a field 20 kpc out on the SE minor axis of M31.

possible to reach some tentative conclusions regarding the mean metallicity of the bulge of M31 and its gradient. The 12 kpc field is the simplest to interpret. There is no excess reddening problem in this field, and no disk contamination. We can be confident of this, because the surface brightness is a lot fainter than 26 mag/s^2, and so we are well beyond any projected disk. It is necessary to deal with the problem of the finite range of our metallicity calibration in Figure 5. We can do this by determining the median color of the distribution, and thus ignoring the fact that we do not know the metallicities of stars at $i = 22.5$ and $g - i > 1.5$. Some of these stars may be Mira variables intruding into this luminosity range from above. The median color in Figure 5 is $g - i = 1.46$, very close to the color for M71, and corresponding to [Fe/H] = -0.5 on the Zinn (1984) scale.

The 5 kpc field is harder to deal with, however. Contamination from the disk of M31 is significant in this field. The blue surface brightness 21' on the minor axis is approximately 24.3 mag/s^2 according to de Vaucouleurs (1958). The line of sight intersects the disk at a radius of 21'/cos(78), i.e. 100', where the surface brightness is 25.4 mag/s^2. The disk population can therefore be expected to contribute approximately one third of the total light. This is probably why the median color at $i = 22.5$ in the 5 kpc field is as blue as $g - i = 1.31$. If we ignore the problem of contamination from younger blue disk stars for a moment, we come across a difficulty with the nonlinear nature of the relation between $g - i$ and [Fe/H].

At present there is no intermediate metallicity cluster with a giant branch observed on the Gunn system. Color transformations from (I, V-I) giant branches (Mould, Da Costa and Kristian 1983) suggest $g - i = 1.06$ at $[Fe/H] = -1.5$. So, neglecting disk contamination, median $[Fe/H]$ in the 5 kpc field is formally -0.9.

Finally, in the 20 kpc field, there is a problem with contamination by foreground stars. From the control field observed by Mould, Da Costa and Kristian, one would predict 4 stars per bin in Figure 8. The giant branch is presumably defined then by the three highest bins in the figure, which yield a median $g - i$ of 1.1 and $[Fe/H] = -1.4$.

From these data one can construct the minor axis metallicity profile of the bulge of M31 (Figure 9). The error bars are dominated by calibration and color zeropoint uncertainties. Even with the addition of a point representing the estimate from integrated light of the mean metallicity of the nucleus of M31 (Mould 1978), it is quite impossible to say at present whether one is looking at a shallow but continuous metallicity gradient, or a constant metallicity spheroid with a metal rich central component. A more clear cut comparison in Figure 9 is the metal deficiency of the globular cluster system relative to the field.

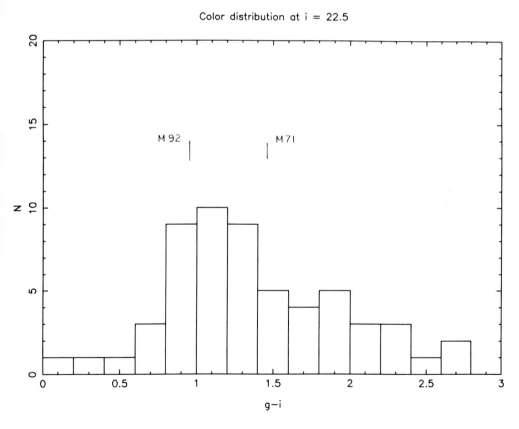

Figure 8. Color distribution on the giant branch in Figure 7.

To estimate the mean metallicity of the clusters, the Mg_2 indices of Huchra and Stauffer (1986) were transformed to the system of Burstein *et al* (1984) by means of 18 objects in common. Burstein *et al* have published Mg_2 indices for 17 Galactic globular clusters, which have metallicities tabulated by Zinn (1985). The first bin consists of all Huchra and Stauffer clusters present on charts 7 and 8 of Hodge's (1981) M31 Atlas, which depict the SE and NW minor axis out to 20'. The mean metallicity of these clusters (101, 108, 114, 150, 178, 213, 222, 230, 231, 233, 234, 235, 250, and 263 of Sargent *et al* 1977) is $[Fe/H] = -1.0$. Their dispersion is 0.6 dex. Identical results are obtained when the published $Mg/H\beta$ or Ca indices of Huchra *et al* (1982) are employed instead. The second bin is made up of all clusters further out on the minor axis . There are only five with data available currently, however (41, 219, 283, 302, and 323). Cluster 41 is also Hubble VIII of NGC 205,

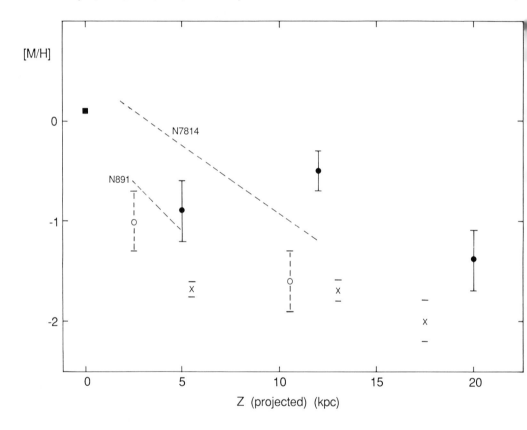

Figure 9. The metallicity profile on the minor axis of M31. Solid symbols: field; open symbols: clusters. The square is an estimate of the metallicity in the nucleus of M31. For comparison the metallicity profile for Galactic globulars is indicated by the crosses (Zinn 1985). The minor axis color gradients of two edge-on galaxies (van der Kruit and Searle 1983) have been converted to metallicity gradients following Aaronson *et al* (1978), and are shown by dashed lines.

but is unlikely to be a member, given its distance from the galaxy and its radial velocity. Counting it in (it does not skew the distribution), we obtain a mean metallicity of [Fe/H] = -1.6, with a dispersion of 0.3 dex. Calcium line indices confirm this result.

The surprise here, of course, is *not* how metal poor the clusters of M31 are. On the contrary, they seem to be more metal rich than those of the Milky Way (also shown in Figure 9 from Zinn 1985 – see also Searle in this volume). Rather, the surprise is how metal rich the bulge is. An order of magnitude higher metallicity at 10 kpc would seem to be significant, even allowing generous uncertainties. A separate study of the mean metallicity in an adjacent field (Mould and Kristian 1986) is consistent with this result at 10 kpc.

The importance of such a result to our understanding of the protogalactic collapse has already received proper emphasis (Strom *et al* 1982) in the context of a very different galaxy, M87. If what we see here is true, the globular cluster system is an earlier population than the field, formed when the metallicity of the protogalactic gas was lower. This is actually a prediction of the theory of cluster formation through thermal instability in infalling metal poor gas by Fall and Rees (1985). Checking the validity of Figure 9 is clearly an urgent matter, both in the immediate sense of obtaining more and better calibrated field and cluster data, and also, in the longer term, of defining the giant branches better and continuing the work into the nucleus. These two are both HST projects. In addition, one needs to construct a BS type model for M31 in order to properly deconvolve disk and bulge. Kinematic studies of the bulge should also be extended beyond 3 kpc, where McElroy's (1984) work stops. Radial velocities for individual giants in M31 will be practical with the new generation of large ground based telescopes.

5. THE CENTRAL COMPONENT OF M31

A central concentration of OH/IR stars has been detected in M31 by Soifer *et al* (1986) at 12μ. In this case, however, the sources are unresolved by IRAS, and we are able to deduce only that this component of M31 has a similar luminosity to that of the IRAS bulge stars in the Milky Way. Optically luminous AGB stars, on the other hand, *can* be resolved close to the center of M31 under exceptional seeing conditions. Figure 10a is a color-magnitude diagram of a 7.8 sq arcmin field 1 kpc out on the SE minor axis of M31 obtained during a fortunate interval at Mt. Palomar of 0.8″ FWHM seeing. Two contiguous pieces of this frame are reproduced in Figure 11, adjacent to a bright star included for identifiability. To preserve the high level of stellar resolution, the measured color in Figure 10 is $i - z$ (and, in passing, we should note that the z magnitudes on this particular night were not well standardized: so the $i - z$ colors may be in error by a fixed additive amount). The point to be made about Figure 10a is that there are rather a large number of stars with $i \simeq 20$, which are not present in the more distant fields of similar area discussed in the preceding section. These stars are clearly rather concentrated to the center of M31, because there are markedly fewer of them in Figure 10b, a field of similar size 1.5 kpc from the nucleus.

Spectroscopy of these stars is necessary to confirm this supposition, but it seems likely that they are the counterparts in M31 of the late M stars found by Blanco, Blanco and McCarthy (1976) in Baade's window. Frogel and Whitford (1986) have shown that i magnitudes alone are unsuitable predictors of the total

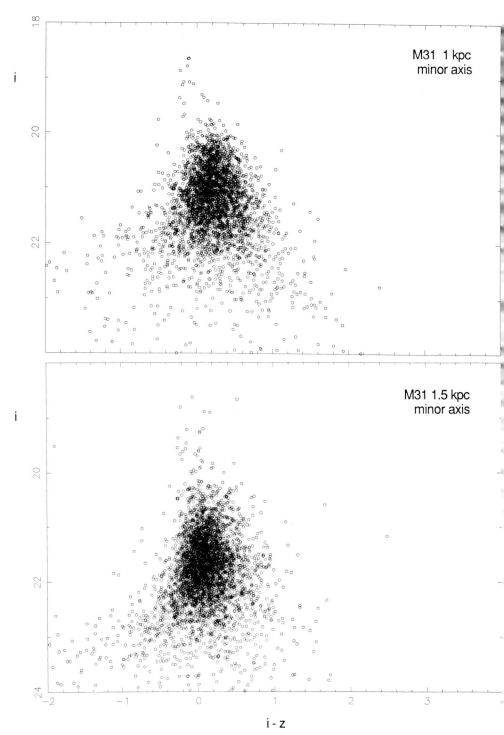

Figure 10. The tip of the giant branch close to the center of M31, *a.* (*abov* 1 kpc, *b.* (*below*) 1.5 kpc on the SE minor axis.

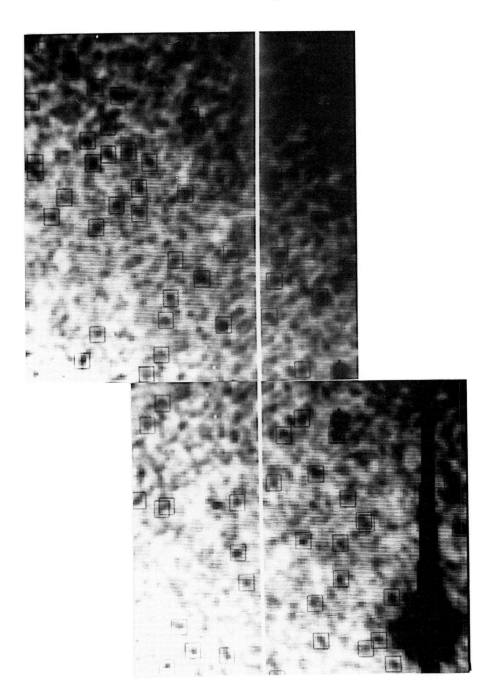

Figure 11. Luminous AGB stars are identified in a field approximately 250″ out on the SE minor axis of M31. The boxes are 6 pixels or 2″ across. A single bad column runs the length of this piece of the 4-SHOOTER image. The saturated star is the bright star 90″ South of G190 on Hodge (1981) chart 1.

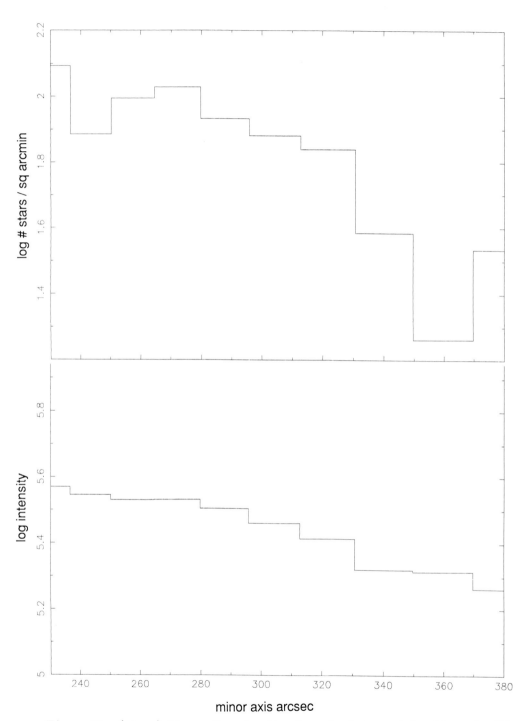

Figure 12a. (*above*) The surface density of resolved stars at $i = 20$ on the SE minor axis of M31; b.(*below*) the unresolved surface intensity in the same region.

luminosities of such cool stars. Adequate measurement will require IR array pho-tometry with large telescopes. But it is already possible to conclude, since $BC_i <$ -0.1 for cluster M giants on the Gunn system, that $M_{bol} < -4.5$ for some of these stars, making them directly comparable with the luminous AGB stars in the Galac-tic nuclear bulge (Frogel and Whitford 1982).

The photometry illustrated in Figures 10 and 12 was carried out by point-spread-function fitting with DAOPHOT. To verify that image crowding, which is extreme in this field (see Figure 11), is not significantly biasing these results, artificial stars were added to the data, located automatically, and measured in the same way as were the real stars. This procedure also allows the true luminosity function to be estimated from the measured one. The corrected luminosity profile for stars with $i = 20 \pm 0.5$ is shown in Figure 12a. To construct Figure 12a, stars were counted in both the real and artificial data between appropriate elliptical isophotes (Kent 1983) superposed on the field. The abscissa of the figure indicates where these isophotes intersect the minor axis of M31. These radial bins are spaced logarithmically so that a power law fit to the profile can be read directly from the figure. It is readily apparent, since the scales are the same, that the $i = 20$ stars are more centrally concentrated than the integrated light at i (Figure 12b). Converted to a volume density distribution, this corresponds approximately to: $\rho(AGB\ stars) \propto r^{-5}$, while $\rho(integrated) \propto r^{-3}$.

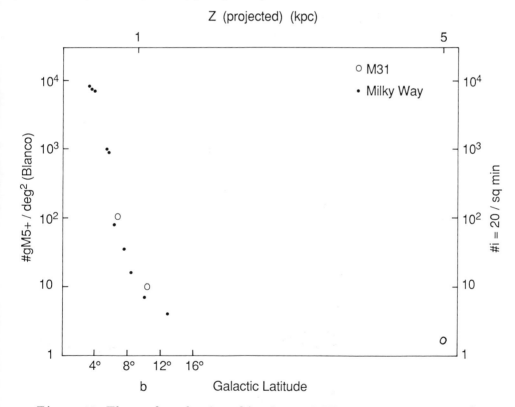

Figure 13. The surface density of luminous AGB stars in the Galaxy (solid symbols, left axis) and in M31 (open symbols, right axis).

On a larger scale the distribution of $i = 20$ stars is compared with the distribution of late M stars in the Milky Way in Figure 13. The three available data points for M31 are superposed on the Galactic data given by Blanco and Blanco (1986). To the extent that $i = 20$ stars correspond to stars later than M5 and that a square arcmin in Andromeda is a square degree in the center of the Galaxy, there is a quantitative similarity in the distributions.

Future work in this area must determine how much more luminous than $M_{bol} = -4.5$ these stars are. Their apparent luminosity function will be a tighter constraint on the radial change in the stellar population towards the centers of galaxies than the corresponding data in the Milky Way, because the distance spread in M31 contributes negligibly. We would like to know if these stars are luminous because they are young, or because they are stingy mass-losers. Both in the Galaxy and in M31, it would be interesting to learn if they are distributed in a spheroid or in a disk. With HST it will be possible to map their distribution all the way into the nucleus of M31. We should also try to map their kinematics as a function of radius.

6. CONSTRAINTS ON THE FORMATION OF BULGES

We characterize stellar populations by their metallicity, age, and kinematics. The data presented here suggest that the bulges of the Milky Way and M31 have a wide range in chemical composition, but are basically metal rich systems, as distinct from the metal poor globular cluster systems. Constraints on the ages of bulges are weak. Clearly they are old stellar populations, but how old relative to globular clusters is an open question. Finally, the kinematic differences between the different systems we have discussed are subtle. The observed velocity dispersions of the globular clusters and the bulge of M31 in the studies cited above are approximately equal at 150 km/sec; and the velocity dispersions of super-metal-rich stars of the bulge of the Milky Way and the globular clusters are *approximately* equal at 110 km/sec.

So where do we look for stronger constraints to tell us how bulges formed ? Some future experiments that come to mind are the following:

1. determine the metallicity distribution function at different radii in the bulge of M31;

2a. determine the star formation history of the inner and outer Galactic spheroid from main sequence stars;

2b. examine the horizontal branch properties of M31 at different radii;

3. enlarge the kinematic samples discussed above to demonstrate real differences.

I want to thank the Space Telescope Science Institute for their hospitality during the preparation of this contribution. Thank you, John Huchra, for making available M31 globular cluster data prior to publication. This work received partial support from NSF grant 85-02518.

REFERENCES

Aaronson, M., Cohen, J., Mould, J. and Malkan, M. 1978; *Astrophys. J.*, **223**, 824.
Bahcall, J. and Casertano, S., 1986, *preprint*.
Bahcall, J. and Soneira, R., 1980, *Astrophys. J. Suppl.*, **44**, 73.

Bahcall, J., Schmidt, M. and Soneira, R., 1983, *Astrophys. J.*,**265**, 730.
Binney, J. 1975, *Phil. Trans. Roy. Soc.*,**00**, 1.
Blanco, V. and Blanco, B. 1986, *Astrophys. and Space Sci.*,**118**, 365.
Blanco, V., Blanco, B. and McCarthy, M. 1984, *Astron. J.*,**89**, 636.
Burstein, D., Faber, S., Gaskell, C.M. and Krumm, N., 1984, *Astrophys. J.*,**287**, 586.
Caldwell, J. and Ostriker, J., 1980, *Astrophys. J.*, **251**, 61.
de Vaucouleurs, G. 1958, *Astrophys. J.*, **128**, 465.
de Vaucouleurs, G. and Pence, W., 1978, *Astron. J.*, **83**, 1163.
Eggen, O., 1983, *Astrophys. J. Suppl.*, **51**, 183.
Fall, S.M. and Rees, M. 1985, *Astrophys. J.*, **298**, 18.
Frogel, J. and Whitford, A., 1982, *Astrophys. J. Lett.*, **259**, L7.
Frogel, J. and Whitford, A., 1986, *preprint*.
Gunn, J., Carr, M., Chang, J., Danielson, J., Lorenz, E., Lucinio, R., Nenow, V., Smith, D., Westphal, J. and Zimmerman, B., 1984, *Bull. American Astron. Soc.*, **17**, 477.
Habing, H., *et al* 1986, *preprint*.
Hartwick, F.D.A., 1984, *Mem. Soc. Astron. Ital.*, **54**, 51.
Hodge, P. 1981, *Atlas of the Andromeda Galaxy*, Univ. of Washington Press: Seattle.
Huchra, J. and Stauffer, J. 1986, *preprint*.
Huchra, J., Stauffer, J. and Van Speybroeck 1982, *Astrophys. J. Lett.*, **259**, L57.
Kent, S. 1983, *Astrophys. J.* **266**, 562.
Kinman, T. 1986, *in preparation*.
Larson, R. 1986, *Mon. Not. Royal Astron. Soc.*, **218**, 409.
McElroy, D. 1984, *Astrophys. J.*, **270**, 485.
Mould, J. 1978, *Astrophys. J.*, **220**, 434.
Mould, J. and Kristian, J. 1986, *Astrophys. J.*, **305**, 59.
Mould, J., Da Costa, G. and Kristian, J. 1983, *Astrophys. J.*, **270**, 471.
Rich, R.M. 1986, *Ph.D. thesis*, Caltech.
Saha, A., 1985, *Astrophys. J.*, **289**, 310.
Sargent, W., Kowal, C., Hodge, P. and van den Bergh, S. 1977, *Astron. J.*, **82**, 947.
Schmidt, M., 1975, *Astrophys. J.*, **202**, 75.
Schneider, D., Hoessel, J. and Gunn, J., 1983, *Astrophys. J.*, **264**, 337.
Soifer, B. T., Rice, W., Mould, J., Gillett, F., Rowan Robinson, M., Habing, H., 1986, *Astrophys. J.*, **304**, 651.
Stetson, P., 1985, *DAOPHOT Users Manual*.
Strom, S., Forte, J.C., Harris, W., Strom, K., Wells, D. and Smith, M. 1981, *Astrophys .J.* **245**, 416.
van den Bergh, S., and Pritchet, C., 1986, *preprint*.
van der Kruit, P. and Searle, L. 1983, *Astron. and Astrophys.*, **110**,79.
Zinn, R. 1985, *Astrophys. J.* **293**, 424.

DISCUSSION

KING: I wish that the word "spheroid" could be expurged from astronomical terminology, because it leads to such confusion. We clearly have a metal-poor halo and a metal-rich bulge, and I see no reason for associating them with each other just because they have similar shapes. We have surely passed beyond the age of Pythagoras and Plato and need not make geometry so sacred that it obscures real astrophysical differences.

MOULD: Your preferred nomenclature is just fine provided we can recover the word "halo" from the dynamicists who use it to refer to dark matter.

WALBORN: Aren't many of your most luminous M31 objects likely to be selectively unresolved multiple images?

MOULD: No, I have done careful simulations to determine and correct for the percentage of objects which are excessively bright because of multiplicity.

RENZINI: Did you try to estimate the lifetime of these bright AGB stars?

MOULD: It is possible to determine lifetimes from the surface densities in Figure 12a. The surface brightness of the galaxy over the region of interest and your very useful "BLT formula." The answer is of order 10^5 years. If the population which produces these stars only contributes a fraction f of the integrated light, however, these lifetime estimates rise of $10^5/f$.

DaCOSTA: Are you prepared to place any limits on the fraction of "intermediate age" stars in your halo fields? For example, in the 12 kpc field there seemed to be a significant number of stars with i ~ 21.5, g - i > 2.0.

MOULD: At i $\approx$ 20 one can make a case for intermediate age stars in the central bulge of M31. I don't think there is an excess over foreground at 12 kpc.

SANDAGE: You show a continuous curve of D(r) vs. r for the galaxy that follows a $r^{-3.5}$ curve. What is the origin of the data that leads to your suggestion that there may be only a single kinematic or dynamical component to the galaxy?

MOULD: That "data" is actually a model which describes the rotation curve of the galaxy. It just looks noisy because the model was differentiated in a simple-minded way. Zinn (1985) has shown that the volume density distribution of globular clusters follows $r^{-3.5}$. Other components may or may not follow different density laws. I don't think we know yet.

SMITH: Your histograms over color for the M31 halo fields show a significant red component, but also a component as blue or bluer than M92. Can you yet say how this blue, presumably metal poor, component compares with that in the halo of our Galaxy?

MOULD: I need to determine the photometric errors before we can tell whether
the spread in the giant branch extends bluer than M92.   In previous work with
DaCosta and Kristian on the dwarf ellipticals around M31, we have not seen any
evidence for a more metal poor population than M92.

GILMORE: The IRAS data base can be used to isolate late M giants from their
cool dust shells. A sample isolated in this way has been constructed and used
to provide a surface brightness map of the Galaxy. Modelling of these data
show the bulge to be very flat, with an axial ratio ~0.45. This provides some
evidence for a discrete central component to the Galaxy. It is, of course,
unclear if the flatness is due to a steep abundance gradient down the minor
axis only, or due to a real flattening of the isodensity contours.

KINEMATICS OF LOCAL SUBDWARFS AND THE COLLAPSE OF THE GALAXY

Allan Sandage

Mount Wilson and Las Campanas Observatories of the
Carnegie Institution of Washington, Pasadena, California

ABSTRACT

The correlations between metallicity and kinematics
of old stars is reexamined using a newly obtained sample
of ~900 stars which, when added to the available
literature data gives a set of 1125 high proper motion
stars. All kinematic properties vary monotonically with
metallicity over the entire range of [Fe/H] from 0 to -4.
The Gilmore-Reid-Wyse thick disk is prominent in the data
with $\langle V \rangle_{LSR}$ = -30 km s^{-1} and $\sigma(W)$ = 42 km s^{-1}. The
rotation of the halo stars, defined as those with $|W| \geq$
60 km s^{-1}, varies monotonically with metallicity from
$\langle V_{rot} \rangle$ = 220 km s^{-1} for $\langle[Fe/H]\rangle$ = 0 to 0 for $\langle[Fe/H]\rangle$ ~
-2.5, indicating spin-up as the galactic collapse
proceeded. The result is not endemic to our kinematically
selected sample. The data are consistent with the ELS
picture of the formation of the Galaxy taking place as a
continuous coherent process rather than by merger of
disparate parts, each with their own enrichment history.

INTRODUCTION

The study of galactic structure from the viewpoint of Stellar
Populations became the dominant theme in work on the Galaxy almost
from the earliest days of the population concept. The concept did not
begin with Baade's (1944) classic paper where many previous results on
kinematics and color-magnitude diagrams were tied into a coherent
overview, sparked by his resolution of the bulge and disk of M31, M32,
and NGC 205. Rather, like the first thin stream of water that
eventually becomes a mighty river, the earliest data, separating what
we now call Populations, were related to velocities of various types
of stars. In the first 20 years of this century, Boss from proper
motions and Adams and Joy from radial velocities discovered parts of
the high and low velocity phenomenon which is so characteristic of the

extremes of the populations. And like a fully flowing river, the population channels have widened to include the tributaries of chemical composition and age, as well as velocity and spatial distribution. These new population concepts now rage through most of stellar astronomy.

The route to the current wider concepts of galactic structure and stellar evolution include the seminal summaries of Strömberg (1924, 1925) of the asymmetric drift, the concept of kinematic subsystems by Lindblad (1925a, b; 1959), the concept of galactic rotation by Lindblad and Oort, the resolution into stars of Shapley's Sculptor and Fornax systems by Baade and Hubble (1939) and the discovery of RR Lyrae variables in Fornax, Baade's resolution of the red stars in M31 and the resulting connection of the resolved stars with the high velocity globular clusters and their associated variables, the finding of the main sequence in globular clusters soon after the 200-inch telescope began operations, the understanding of the HR diagram in terms of evolution off the main sequence, and the subsequent division at the Vatican Conference into five Lindblad kinematic subsystems of parts of the Galaxy, partially on the basis of the new evolutionary concepts.

It would indeed be wrong to believe that the five grades of stellar populations in the Galaxy that were set out by the Vatican Conference (O'Connell 1958) are to be interpreted that the Galaxy is so digitized. Rather, the classification embodied by the scheme represents stages of what is, in fact, a continuum that itself has structure. It has features of a "discontinuous continuum"--a continuum whose gradients of properties change with position in the Galaxy in such a way that it is convenient to characterize the system as if the components were real. Following the Vatican Conference, these components are now called the young thin disk, the old thin disk, the old thick disk, the intermediate halo, and the extreme halo populations. This description is the modern version of the Lindblad kinematic subsystem concept because each of the components has semi-discrete kinematic and spatial properties within the Galaxy.

The language of the Vatican Conference is still useful. Again, its use does not imprint a fictional discretization on the Galaxy as claimed by the opponents of the ideas, ideas which when taken to their limits merely identify the extremes of the modern population concept to be Baade's original separate populations I and II; or, as we would say today, the young thin disk and the extreme halo components of galactic structure.

I shall write this report in the language of the Vatican Conference. Therefore, throughout this review we shall talk of the disk and the halo populations, and shall further be forced to divide the disk into two components--the thin disk and thick disk--on the basis of almost discrete kinematic properties of certain samples of stars, following the work of Gilmore and Reid (1983), Gilmore and Wyse (1985), and Wyse and Gilmore (1986).

THE NEW KINEMATIC SAMPLE

It became evident through Roman's (1954, 1955) work that the very small percentage of stars in the Galaxy with high velocity (Strömberg 1925; Oort 1926) have peculiar spectra--the metal lines are weak relative to the hydrogen lines at a given temperature. In the 1950's these subdwarf stars would soon be identified with the globular cluster main sequence stars by a variety of tests (Sandage 1986a for a review).

With the data available in the early 1960's, Eggen et al. (1962, hereafter ELS) found correlations between various kinematic properties and the metal abundance. These involved relationships between [Fe/H] and the eccentricity e of the projected (planar) orbit, the angular momentum h, and the W velocity for any given star. From the available data, ELS concluded that the kinematic and metallicity correlations were evidence that the formation of the Galaxy occurred by collapse from a larger volume, with progressive metal enrichment as the collapse proceeded, leading to the formation of the disk by gas-gas collisions (dissipation).

The number of stars with [Fe/H] < -1.4 available to ELS was only ~50. This range of metallicity is the domain which is believed to contain most of the clues to the early formation history of the Galaxy.

To increase the ELS sample, various astronomers began programs in the 1960's to find more such stars. The Bruce proper motion survey (Luyten) and the Lowell proper motion survey (with a team led by Giclas) were often used as the source for candidates. As one of these subdwarf discovery programs, we began at Mount Wilson an observational campaign with Kowal in 1960 to obtain photometric data for ~1700 high proper motion stars (Sandage and Kowal 1986), and with Fouts in 1980 (Fouts and Sandage 1986) to obtain radial velocities for ~900 of these. The data, combined with ~200 high velocity stars in the literature gives a sample of 1125 stars with which we have been able to study again the correlations found by ELS between kinematics and chemical composition.

A parallel investigation was begun by Carney and Latham in ~1980 with a comparable number of stars. The Carney-Latham data and analysis are not yet available, and their addition to the sample

discussed here should eventually improve the results. The remainder
of this review will be concerned with the consequences of the new data
obtained with Kowal and with Fouts.

RESULTS

    The detailed analysis of the new sample has been set out in a
paper for the archival literature (Sandage and Fouts 1987, Paper VI).
The purpose of the present report is to summarize the evidence and to
discuss the meaning of the data as they relate to the ELS view of how
the Galaxy formed and evolved into its present system (either discrete
or as a "discontinuous continuum"). The summary is most conveniently
given by the following ten points.

    (1) The U, V, and W Galactic velocity components for the 1125
final sample, displayed in various parameter spaces that involve the
kinematics and the chemical compositions (measured by the ultraviolet
excess), show the presence of two kinematic components. These are the
halo and the Gilmore-Reid-Wyse thick disk with a Strömberg
asymmetric drift of only $\langle V \rangle = -30$ km s^{-1} (relative to the local
standard of rest), and $\sigma(W) = 42$ km s^{-1} corresponding to a scale
height of 920 pc. The various representations in which we can
identify the two components include the Bottlinger U,V diagram, the
V,δ diagram, the V,W diagram as a function of δ, the e,δ, the e,h, and
the U,δ diagrams.

    The Bottlinger diagram for four different ultraviolet excess bins
is shown in Figure 1 for the total sample. The δ bins correspond to
metallicity values of $\langle [Fe/H] \rangle = 0, -0.6, -1.1,$ and -2.5 respectively.
The progressive change in the nature of the distribution as the
metallicity decreases is the principal feature of Figure 1.

    The mean height above the plane for each of the distributions in
Figure 1 can be obtained from the W velocities as they change with δ
over the Bottlinger diagram. It is this binning of the data by W that
betrays the fact that the clump of low U,V stars in the upper two
panels of Figure 1 has a low $\sigma(W)$ compared with the bottom two
panels, identifying the thick disk component. The evidence is shown
in Figure 2 which is the V,W space for various δ values. The large
change in $\sigma(W)$ as [Fe/H] decreases from 0 to ~ -4 is clear. The
thick disk component is confined to $\delta \leq 0.15$ and $V \leq -100$ km s^{-1}.
The mean drift velocity is $\langle V \rangle = -45$ km s^{-1} relative to the sun.
Hence the system lags the rotation of the local standard of rest but
not by very much. It may, therefore, represent the rapidly rotating
bulge component in the solar neighborhood, similar to bulges in S0
galaxies (Dressler and Sandage 1983).

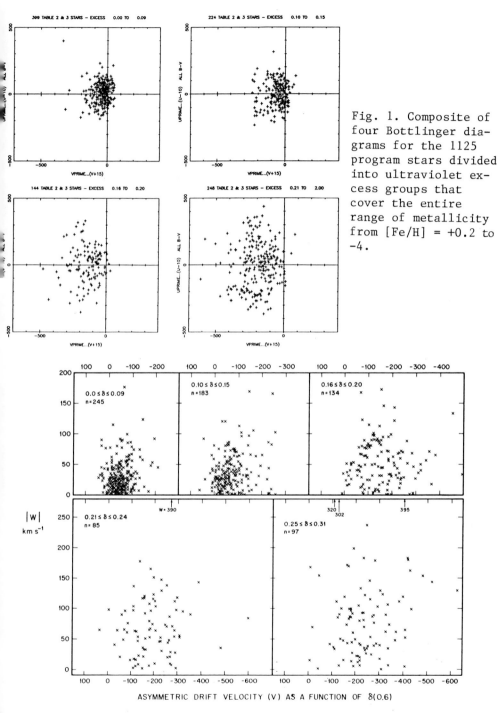

Fig. 1. Composite of four Bottlinger diagrams for the 1125 program stars divided into ultraviolet excess groups that cover the entire range of metallicity from [Fe/H] = +0.2 to -4.

ASYMMETRIC DRIFT VELOCITY (V) AS A FUNCTION OF δ(0.6)

Figure 2.   The V,W diagram for five intervals of δ for stars in the sample with B-V ≦ 0.8. The velocities are heliocentric, not reduced to the local standard of rest.

The metallicity distributions of the thick disk and the halo, obtained from the distribution of δ values in Figure 2, are shown in Figure 3. The thick disk has [Fe/H] intermediate between the halo and the thin disk as defined, for example, by Twarog's (1980) ~1000 star solar neighborhood sample, or the trigonometric parallax sample (Gliese 1969).

(2) Within the thick disk itself there is a metallicity gradient with height, shown by the wedge-shaped distribution of the W,δ diagram in Figure 4 for the thick disk stars alone, defined as those with $V >$ -100 km s^{-1}. Note from this diagram that $\sigma(W)$ progressively increases with increasing ultraviolet excess. This shows that the scale height within the thick disk increases with decreasing metallicity, requiring a metallicity gradient in situ.

Within the halo component alone there is a monotonic variation of both the $\sigma(W)$ values and the asymmetric drift velocity $\langle V \rangle$ with [Fe/H] over the whole of the metallicity range from +0.2 to -4. The evidence on the continuous variation of the asymmetric drift is seen in Figures 1 and 2 and more clearly later in Figures 7 and 8. The variation of $\sigma(W)$ with [Fe/H] is evident in Figure 2 for halo stars, and is more clearly seen from the W,δ diagram, discussed as the next item.

(4) The large range of [Fe/H] for halo stars in Figure 3 is not present at every height but decreases markedly with increasing $|W|$. Figure 5 shows the W,δ diagram for halo stars alone; the thick disk stars have been removed by restricting the sample of $V < -100$ km s^{-1}. The distribution is wedge-shaped, similar to Figure 5 of ELS (but now containing only halo stars). The histograms of the W distributions in various bins of δ are shown in Figure 6, showing in a different representation the progressive increase of $\sigma(W)$ with decreasing metallicity. As in ELS, this is interpreted to mean that very low metallicity stars can form at any height but the mean height of formation decreases with increasing metallicity. As in ELS, the postulated reason is that because [Fe/H] increases with time, stars formed later will have higher [Fe/H] values on the average. Because, from Figure 5 such stars have progressively smaller $|W|_{max}$ values, hence smaller heights, collapse of the parent gas distribution is evident. However, to explain Figures 5 and 6 there must have been gas-gas dissipation (even at these earliest epochs) within the halo gas at heights corresponding to $|W| \sim 300$ km s^{-1} or ~30 kpc so as to prevent the high [Fe/H] stars (formed at lower heights) from rising to the large starting height of the outermost region of the original gas cloud. Otherwise, Figure 5 would be box-shaped rather than wedge-shaped.

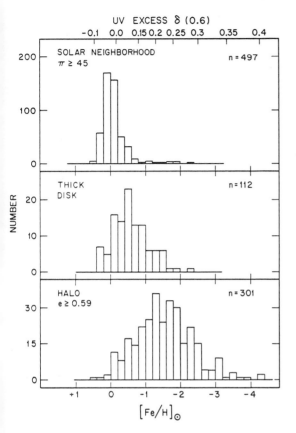

Fig. 3. Histograms of the metallicity distributions of the old thin disk, the old thick disk, and the halo as inferred from the distributions of δ. The data for the old thin disk are from the Gliese (1969) trigonometric parallax catalog for stars with $0.3 \leq B-V \leq 1.0$ and $\pi \geq 0\overset{''}{.}045$.

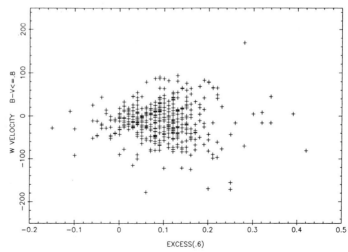

Fig. 4. The correlation of W with δ for thick disk stars defined by V > -100. Only stars with $B-V \leq 0.8$ are used.

(5) The distribution in the W,δ diagram of Figure 5 is not affected by the observational bias in our kinematically selected sample which misses the low velocity stars. The empty region for δ < 0.1 (i.e., the high metallicity domain) are the <u>high</u> velocity parts of the diagram ($|W| \geq 100$ km s^{-1}) for which our sample is optimized to study. The lack of many high metallicity stars with high W is therefore real in our sample.

(6) The increasing σ(W) values for increasing δ shown in Figure 6 requires the scale hight for halo stars to change from ~1 kpc in the metallicity range of $0 \geq [Fe/H] \geq -0.6$ to ~5 kpc for the metallicity range of $-1.8 \geq [Fe/H] \geq -4$. This again requires a chemical gradient in the high halo.

(7) The V,δ diagram for halo stars alone in Figure 7 shows that the Strömberg asymmetric drift velocity increases monotonically with δ. The mean values of <V> at given <δ>, determined from the data in Figure 7, are shown in Figure 8 as the galactic rotation (220-<V>$_{drift}$) vs <[Fe/H]> determined from the <δ> values. The monotonic decrease of v_{rot} with decreased metallicity suggests that spin-up has occurred for halo stars as the collapse proceeds. This conclusion is opposite from that of Norris (1986) who states that his data show no correlation of kinematics with metallicity for stars with [Fe/H] < -1.2. Norris' data on the galactic rotation are shown in Figure 8 as triangles. The agreement of his data with ours is excellent except for the last metallicity bin. There we show a decreased rotational velocity rather than a flat distribution in this high velocity bin where our sample is optimal for discovery.

(8) The difference between our conclusion and that of Norris is not because of a bias in our sample due to its kinematic selection. The trend in Figure 7 is toward <u>higher</u> velocities with decreasing [Fe/H]. Again, our sample is optimal for discovery of the high velocity stars. To counter the incresed <V> with δ for δ ≥ 0.16 in Figure 7 we must be missing enough stars in the semiblank region of $|V| < 100$ km s^{-1} and δ > 0.24 of the diagram to keep <V> constant over the entire range from 0.16 ≤ δ ≤ 0.32 if Norris, and Searle and Zinn (1978) are correct. A stronger argument that the progressive change of v_{rot} with [Fe/H] in Figure 8 is correct is the observed increase of σ(U) with δ, shown by our data in Figure 9. These follow the required Strömberg-Jeans <V(lag)> ~ $[σ(U)]^2$ relation. If there was in fact no correlation of V(rot) with [Fe/H] for [Fe/H] < -1.2, the correlation in Figure 9 would not exist.

(9) From the circles in the V,$(U^2+W^2)^{1/2}$ Toomre energy diagram of Figure 10 that encompass all the points, we can estimate the escape velocity from the Galaxy at the solar circle. The total mass of the Galaxy follows from the flat rotation curve approximation that gives

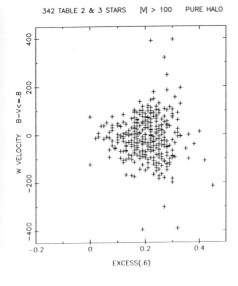

342 TABLE 2 & 3 STARS   M > 100   PURE HALO

Fig. 5.   The correlation of W with δ for halo stars in the sample with B–V ≤ 0.8.   The thick disk component has been removed by the kinematic criterion V < -100 km s⁻¹.

Fig. 6. Histograms of the W distribution for different intervals of δ for the halo stars in Fig. 4.

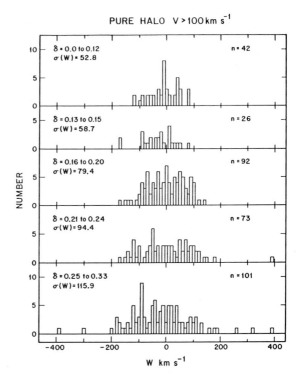

$$V^2_{E,R_\odot} = 2V^2_{rot,R_\odot} [1 + \ln M_T/M_{R_\odot}]$$

where V_{rot} is the rotation of the Galaxy at the solar circle $R_\odot$, $V_{E,R_\odot}$ is the escape velocity, $M_{,R_\odot}$ is the mass of the Galaxy inside the solar circle, and M_T is the total mass. Our sample was not selected to find the highest velocity stars. Hence, the envelope circle near $V_{E,R_\odot} = 450$ km s^{-1} is a minimum value. The deeper sample of Carney and Latham suggest a value of $V_{E,R_\odot} = 550$ km s^{-1} which requires eight times more mass outside the solar circle than within it (assuming $V_{rot,R_\odot} = 220$ km s^{-1}). The high value of $V_{E,R_\odot} = 650$ km s^{-1} advocated by Caldwell and Ostriker (1981) requires $M_T/M_{R_\odot} = 29$, or M_T(Galaxy) $= 2.5 \times 10^{12}$ which, when added to the nominal M31 mass (as double the Galaxy), gives M_T(Local Group) $= 7.5 \times 10^{12}$ M$\odot$. This is well beyond the limit set by the cosmological deceleration test for the Local Group (Sandage 1986b). From this argument we consider $V_{E,R_\odot} = 650$ km s^{-1} to be impossibly high.

(10) The principal new results compared with ELS are (a) the presence of the strong Gilmore-Reid-Wyse thick disk component of intermediate metallicity in our kinematically selected sample, (b) a strengthening of the evidence for a W,δ correlation for halo stars alone, requiring a chemical gradient in the high halo, and (c) the progressive change of $\langle V \rangle_{lag}$ with increasing δ, suggesting spin-up with increasing collapse factor.

We can only conjecture how the thick disk formed, keeping within the general outline of the ELS picture. The intermediate [Fe/H] value for the thick disk is perhaps the central clue to its formation. We conjecture that the strength of the gradient, and consequently, the mean value of [Fe/H] at any height depends on the ratio of the rate of dissipation (i.e., decay rate of the Z energy of position of the parent gas clouds) compared with the rate of metal enrichment. If there is no dissipation but only free-fall, there will be no gradient, as emphasized by Searle and Zinn (1978).

Because there is a slight gradient in the high halo for $\delta \gtrsim 0.16$ (Figure 5) and a wide distribution of [Fe/H], but a much more collapsed spatial structure and a higher $\langle$[Fe/H]$\rangle$ for the thick disk, we conjecture that a nonlinear dissipation process took place--the closer to the plane the parent gas had plunged, the higher became pressure support and therefore the greater the dissipation rate, and hence the larger was [Fe/H] at a given height late in the collapse pahse. To produce a discrete structure such as a true thick disk separated from the halo phase requires a discontinuity in the dissipation rate, occurring when the maximum scale height of the parent gas out of which the thick disk formed was at ~1 kpc. One can suggest that at the time when the collapse has proceeded to ~1 kpc height above the structure that was becoming the galactic plane, the

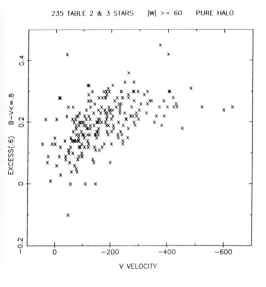

235 TABLE 2 & 3 STARS |W| >= 60   PURE HALO

Fig. 7. The correlation of V with δ for halo stars alone, the thick stars being removed using the kinematic criterion of $|W| \geq 60$ km s^{-1} to define the halo. The sample is restricted in color to B–V $\leq$ 0.8.

Fig. 8. Variation of V_{rot} with [Fe/H] for halo stars in our sample (dots), for the thick disk (the cross), and for Norris' (1986) sample (triangles) of stars chosen spectroscopically to represent the "halo."

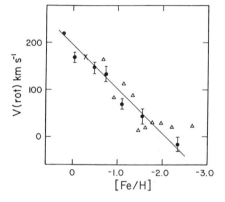

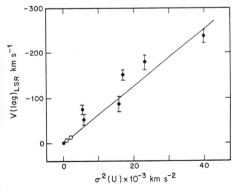

Fig. 9. The Strömberg asymmetric drift velocity as a function of $[\sigma(U)]^2$ for the halo stars in our sample used in Fig. 7, for solar neighborhood data (open circles), and for the local standard of rest (closed circles).

collapse rate slowed <u>relative</u> <u>to</u> <u>the</u> <u>metal</u> <u>enrichment</u> <u>rate</u> (due to
increased pressure support), permitting higher values of [Fe/H] at any
height Z ≤ 1 kpc than would have been obtained if the dissipation
rate had remained constant. The physics of the cooling rate is, of
course, now the unaddressed issue.

This report summarizes the last phases of the Mount Wilson Halo
Mapping Project which has been supported by the U.S. National Science
Foundation grant AST 82-15063 in the last three years for which I am
grateful. It is also a pleasure to thank Gary Fouts for his unfailing
efforts in all phases of obtaining and analyzing the new data sample
from 1980 to 1985.

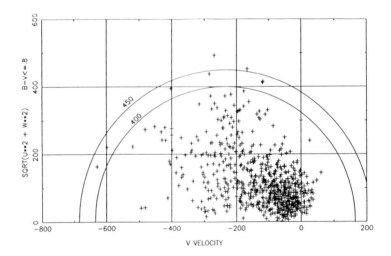

Fig. 10. The Toomre energy diagram for the total sample of stars with
B-V ≤ 0.8. The V velocity is heliocentric not reduced to the LSR.
The two circles are centered on V = -235 km s^{-1} which is assumed to be
the heliocentric velocity of rotation.

REFERENCES

Baade, W. (1944). Astrophys. J. 100, 137.
Baade, W. and Hubble, E. (1939). Publ. Astron. Soc. Pac. 51 40.
Caldwell, J.A.R. and Ostriker, J. P. (1981). Astrophys. J. 251, 61.
Dressler, A. and Sandage, A. (1983). Astrophys. J. 265, 664.
Eggen, O. J., Lynden-Bell, D., and Sandge, A. (1962). Astrophys. J.
    136, 748.
Fouts, G. and Sandage, A. (1986). Astron. J. 91 1189.
Gilmore, G. and Reid, I. N. (1983). Mon. Not. Roy. Astron. Soc. 202,
    1025.
Gilmore, G. and Wyse, R.F.G. (1985). Astron. J. 90, 2015.
Gliese, W. (1969). Veröffentichunger des Astronomishen
    Rechen-Instituts, Heidelberg, Nr. 22.
Lindblad, B. (1925a). Upsala Medd. No. 3.
Lindblad, B. (1925b). Ark. f. Mat. Astron. Och. Fysik 19a: No. 21.
Lindblad, B. (1959). In Handbuch der Physik, ed. S. Flügge, p. 21.
    Berlin: Spinger-Verlag, Vol. 53.
Norris, J. (1986). Astrophys. J. Suppl., in press.
O'Connell, D.J.K. (1958). Ric. Astron. Specola Vaticana 5 (Stellar
    Populations).
Oort, J. H. (1926). Publ. Kapteyn Astron. Lab., Groningen, No. 40.
Roman, N. G. (1954). Astron. J. 59, 307.
Roman, N. G. (1955). Astrophys. J. Suppl. 3, 195.
Sandage, A. (1986a). Annual Rev. Astron. Astrophys. 24, in press.
Sandage, A. (1986b). Astrophys. J., in press.
Sandage, A. and Fouts, G. (1987). Astron. J., submitted (Paper VI).
Sandage, A. and Kowal, C. T. (1986). Astron. J. 91, 1140.
Searle, L. and Zinn, R. (1978). Astrophys. J. 225, 357.
Strömberg, G. (1924). Astrophys. J. 59, 228.
Strömberg, G. (1925). Astrophys. J. 61, 363.
Twarog, B. A. (1980). Astrophys. J. 242, 242.
Wyse, R.F.G. and Gilmore, G. (1986). Astron. J. 91, 855.

DISCUSSION

KOO: (1) The fraction of binaries among halo stars remains
controversial, but if it is as high as 20-30% as recently suggested by
e.g., Stryker et al., to what extent will such binaries affect your
photometric distance estimates?
(2) Does the thick disk component have the same LF as that of the halo
or old disk?

SANDAGE: The answer to the first question is given by Laird in the
comment that follows this. I do not know the answer to the second
question. Our sample is confined to B-V $\leq$ 1.0, and further is highly
restricted by the PM cutoff, so it is not very well-suited to derive
the luminosity functions. However, by studying the photometric
parallaxes of high PM type K and M stars, it would be possible to test
if the thick disk component contains late K and M stars with $\sigma(W) \simeq$
40 km s^{-1} for example, as a discrete population. I have a large
amount of data on photometric parallaxes of M high PM stars, and will
look into your suggestion.

LAIRD: I'd like to make a comment about the binary frequency based on
the Carney-Latham survey. They find a binary fraction of about 15-20%
from velocity variations. The number of double-lined systems, for
which there is direct evidence for problems with visual colors, is
very much smaller. So, the photometric parallaxes are generally
probable reliable, while the binary fraction is still fairly high.

van den BERGH: Could it be that the discontinuity at $\delta(U-B) \approx 0.14$
corresponds to the point where galactic collapse had progressed to the
point where stars could no longer be formed with retrograde orbits?

SANDAGE: That is certainly an interesting point, supported by Figure 1
of the text.

BURSTEIN: (1) What are the mean distances of your stars?
(1) Although I realize there are good reasons to think the stars you
are looking at are free of local bias, is it possible the localized
phenomena, like Heiles' superbubbles could affect these data over
~5 x 10^9 years?

SANDAGE: (1) The bulk of the stars are within 250 pc, with a mean of
D ~ 70 pc (Figure 9 of Astron. J. 91, 1182, 1986).
(2) If the question is on the kinematics of the orbits, such effects,
if they exist, would only increase the scatter of the correlations
between metallicity and orbital properties. Because the systematics
seem moderately well-behaved as is, the true situation must then be
even tighter.

MOULD: Can you comment on the possibililty that the intermediate
velocity population seen in your data may be a flattened spheroid
rather than a thick disk?

SANDAGE: I have always thought the two descriptions to be
interchangeable. I believe your suggestion fits the fact that the
"flattened spheroids" of S0 galaxies are rapidly rotating (Astrophys.
J. 265, 664, 1983) just as is the $\langle V \rangle_{LSR}$ = -30 km s^{-1}, $\sigma(W)$ = 42 km
s^{-1} structure identified here as the Gilmore-Reid-Wyse "thick disk."

KING: Since your velocities come from proper motions and photometric
parallaxes, the errors in the parallaxes will cause errors in the
velocities. I think that this will go in the direction of making the
velocities look somewhat larger than they should.

SANDAGE: Yes, from the standpoint of the formal errors, this is the
direction of a small effect. A much more serious problem is that we
have assumed all the stars to be on the main sequence, whereas some
~15% will be evolved subgiants just beyond the turnoff. Our assumed
main sequence photometric distances are, then too small for them, and
their estimated velocities are therefore also too small. Hence, our
velocity diagrams show mostly the lower limit possibilities.

KING to FREEMAN: If the UV-excess errors were smearing things out,
then Sandage would not have the sharp discontinuity that he showed at
δ = 0.15.

PIER: What does the velocity ellipsoid look like for your sample? In
particular, what is $\sigma(U):\sigma(V):\sigma(W)$? Does it change with/without
your "thick disk" component? Does it change with $\delta(U-B)_{0.6}$?

SANDAGE: The total sample shows:

δ	$\sigma(U)$	$\sigma(\lvert V-\langle V \rangle \rvert)$	$\sigma(W)$
<0.01	59	29	27
0.00 to 0.09	67	42	36
0.10 to 0.15	93	61	47
0.16 to 0.20	128	94	73
0.21 to 0.33	173	107	104

For the thick disk, the mean values are $\sigma(U)$ = 80, $\sigma(V-\langle V \rangle)$ = 40, and
$\sigma(W)$ = 42 but with some variation with δ. For the halo, the variation
with δ is monotonic. Values for $\langle \delta \rangle$ = 0.23 are $\sigma(U):\sigma(V):\sigma(W)$ =
154:87:94.

LARSON: I was impressed with the evidence for a continuous collapse of
the halo, presumably reflecting a progressive shrinkage, spinning up,

and enrichment of a gas cloud. However, I'm puzzled about why evidence
for this hasn't shown up previously, for example, in studies of
globular clusters. Is it just a matter of sample size? Or is there a
real discordance between the properties of halo field stars and those
of globular clusters?

SANDAGE: The globular cluster sample _is_ very small and one only has
the radial component to their velocities. However, spatially, the 47
Tuc like metallicity clusters do have a scale height of ~1 kpc.

As to the spin-up, ELS and later more extensive subdwarf data do
contain the monotonic variation of the asymmetric drift velocity with
[Fe/H] over the whole of the [Fe/H] range. Pier's data on the HB blue
stars also shows a difference in the rotation of his two metallicity
groups in the direction we have obtained here.

LOCAL GROUP DWARF GALAXIES: THE RED STELLAR POPULATION

Marc Aaronson
Steward Observatory, University of Arizona, Tucson, Arizona

Abstract. We summarize recent studies of the red (i.e. older) stars in nearby dwarf systems, and the implications this work has with regard to the origin, evolution, and interrelationships of differing types of low-luminosity galaxies. It has become clear that dwarf irregulars and dwarf spheroidals share many common properties, circumstantially suggesting that these two kinds of objects share a similar heredity. The mass-to-light ratios of both dI and dSph systems appear to exhibit a steep dependence on absolute magnitude, in good accord with recent models of biased galaxy formation with cold dark matter. Alternatively, the observations may indicate that the initial mass function depends sharply on abundance.

1. INTRODUCTION

We begin by addressing the question of why one should study dwarf systems, the puniest members of the galaxian zoo. One reason is their apparent structural simplicity, which ought to make them ideal for testing various evolutionary models. A second reason involves the extreme properties exhibited by some dwarf objects, as it is often the case that we can gain considerable insight into a class as a whole by studying the most deviant members. Democracy of course demands that some attention be paid to dwarfs, since they appear to be the most numerous types of galaxies in existence. This last point has now taken on a rather special significance in view of the increasing likelihood that dwarfs possess substantial mass-to-light ratios. Finally, a certain degree of satisfaction accompanies study of the wide variety of dwarfs we are lucky to be surrounded by in the Local Group. In particular, and especially in view of the advances in astronomical technology made in recent years, we can examine the population of nearby dwarfs on a star-by-star basis, and avoid the frustrations inherent in composite light measurements of more distant objects.

The organizers of this conference have invited me to discuss the red stellar population of dwarf galaxies, which I have interpreted really to mean the older stars -- those with ages greater than ~ 1 Gyr and masses less than ~ 2 $M_\odot$. This stellar component is of course key for understanding the star formation and chemical enrichment histories of galaxies. It also provides critical clues to the origin and evolutionary relations of the differing dwarf types. The latter range from the spheroidals (dSph) -- dwarfs devoid of H I gas and dead from the standpoint of recent star formation; to their somewhat larger cousins, the dwarf ellipticals (dE), which in some cases exhibit residual star formation right up to today; through the dwarf irregulars (dI) -- systems undergoing a moderate degree of present day star formation; to the blue compact dwarfs (BCD) -- objects suggested by gas consumption arguments to be undergoing intense but short-lived star bursts. Although not the topic here, there is obviously much to be learned about the physical processes of forming stars from studying dwarf galaxies. Indeed, the principal distinguishing feature among dwarfs -- the ratio of present to past star formation rate -- seems far less tied to the overall global morphology than in "normal" size galaxies, which seemingly demands the operation of some sort of stochastic processes at work. One final aspect of the study of older stars in nearby dwarfs that is worth mentioning involves the feedback provided to stellar evolutionary models. The last few years have seen an especially rewarding symbiosis of observations and theory of asymptotic giant branch (AGB) stars.

In the review we shall only discuss observations of nearby resolvable systems. This necessarily leaves out the BCD's, which are covered by W. Sargent in another article in this volume. We also forego talk about the many new composite light studies of Virgo cluster dE and dI systems, whose impetus has been provided by several recently completed large-scale dwarf surveys. Suffice it to say that from the viewpoint of the author, these other results seem reasonably consistent with our knowledge of the stellar populations in the nearer dwarfs.

The article is divided into three parts. We first consider the Magellanic Clouds (Section II), with emphasis on what has been learned about the enrichment rate and strength of past star formation bursts. We next turn to the other dI and dE systems in the Local Group, focusing on the question of how perturbed or unusual the stellar content of the Clouds is because of their location in the Milky Way's tidal field. Finally (Section III), we give an extensive discussion of the properties of the dwarf spheroidals, devoting some attention to questions involving the presence of non-luminous matter and the origins of these systems. In all of this we shall concentrate primarily on newer results; several recent summary papers on these topics, along with references to older source material, can be found in Frogel (1984), Zinn (1985), and Aaronson (1986).

2.   THE MAGELLANIC CLOUDS

In the crudest terms, we picture the Milky Way to consist of two major populations. Similarly, the Clouds are often characterized by three components: 1) a young (< 1 Gyr) population that includes supergiants, Cepheids, and rich star clusters; 2) an intermediate-age (1-10 Gyr) population signified by field AGB stars (particularly large numbers of carbon stars), as well as rich star clusters again; and 3) an older (> 10 Gyr) population represented by RR Lyraes and a handful of galactic globular-like clusters. We shall not even attempt to summarize the large body of work in this area (for which instead the reader is referred to the IAU Symposium volume, No. 108, on the Structure and Evolution of the Magellanic Clouds), but will instead confine the remarks to the two topics mentioned above.

We start, however, by noting some important population differences between the Clouds and the Milky Way. For one thing, the former have been able throughout their lifetimes to manufacture and keep intact massive star clusters. Interestingly, the luminosity function of these clusters exhibits the monotonic increase seen for Galactic open clusters, rather than the guassian-like peak found for globulars in both the Milky Way and M31 (Elson and Fall 1985a). Second, the carbon-to-M giant star ratio is substantially larger in the LMC than in the Milky Way, and larger still in the SMC (Blanco and McCarthy 1983). Third, perhaps the most intriguing difference is that as far as anyone can tell, the three population components are well mixed both spatially and kinematically (e.g. Freeman et al. 1983).

What is possibly most remarkable about the spatial homogeneity of the Cloud populations is the very large radial distance out to which it extends. To illustrate this point consider the three LMC clusters NGC 2257, ESO 121-SC03, and E2, which all lie a few degrees from each other at a mean distance from the optical center of ~ 9°, i.e. well outside the 5° limit of the Hodge and Wright (1967) atlas. In NGC 2257, the blue horizontal branch, the presence of RR Lyraes, and the position of the main sequence turnoff all point toward a basic similarity with galactic globulars and a corresponding age of ~ 16 Gyr (Hesser et al. 1984). In contrast, the turnoff age for ESO 121-SC03 is ~ 8-10 Gyr (Mateo et al. 1986), while that for E2 is only ~ 1-2 Gyr (Schommer et al. 1986)! Note that all three clusters lie beyond the limits of presently detected H I gas (see Mathewson and Ford 1984). Olszewski et al. (1986) have recently assembled a catalog of ~ 140 LMC clusters lying outside the boundaries of the Hodge and Wright atlas; detailed study of the age-abundance relation and the dynamics of these outlyers is of obvious interest for pinning down the formation history of the LMC halo.

2.1 Enrichment

The star clusters of the Clouds provide an unsurpassed opportunity for making a detailed mapping of the chemical enrichment rate in external systems. The widespread use of solid state area detectors in combination with the development of sophisticated software routines for handling crowded field photometry has led to nothing less than a full Renaissance in this area. While the older painstaking photographic work could barely reach to V ~ 21, it is now routine to obtain CM diagram reaching to V ~ 24 or even deeper in only a few hours of good seeing with a CCD at the prime focus of a 4-m class telescope; indeed, virtually no month goes by without the appearance of a Cloud cluster CM diagram in some journal.

  With all this new material, the age-abundance relation in the Clouds is beginning to take shape, albeit in a still very schematic fashion. Figure 1 shows the emerging LMC and SMC enrichment rates, taken from a recent preprint by Smith and Stryker (1986); also illustrated is the Galactic relation, adopted from Twarog (1980). It is clear from the figure, as has long been thought, that the chemical evolution of the Milky Way is very different from that in the Clouds. In the former, the period of rapid enrichment that presumably coincided with the collapse of the halo was followed by a very slow abundance increase in the disk. In contrast, the chemical build up in the Clouds seems to have occurred at a much more leisurely pace. There is furthermore an indication that the LMC and SMC relations differ from each other, with an even more protracted enrichment timescale having occurred in the latter.

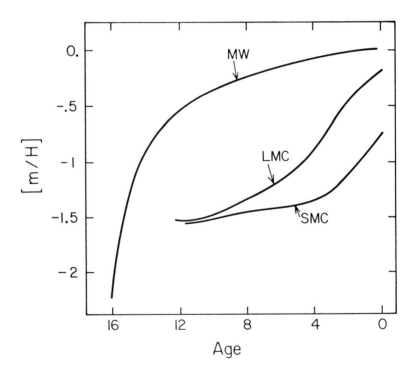

Figure 1. Enrichment history of three galaxies. The relation for the Milky Way is adopted from Twarog 1980, while that for the Clouds comes from Smith and Stryker 1986.

  The so-called "simple" or closed-box model of chemical evolution is to first order thought to provide an adequate description of the enrichment history in Magellanic-type irregulars (Pagel 1981). The simple model combined with a constant star formation rate does appear to provide a reasonable fit to the LMC relation in Figure 1 (see Mould and Aaronson 1982). If the simple model is also to fit the SMC, the formation rate of stars providing the chemical yield may be required to be slower than it was in the LMC. It is interesting to contrast this possibility with the fact that the background stellar sheet in the SMC appears to be of an older average age than in the LMC (cf. Bruck et al. 1985, Hardy and Durand 1984). This difference also seems to show up in the cluster age distributions. Some five SMC clusters (L1, L113, K3, N121, and N339) are now known via the turnoff position to lie within the age range of ~ 5-10 Gyr, while the only LMC cluster so far found within this same range is the aforementioned ESO 121-SC03, a dichotomy which at this point seems hard to account for by some selection effect. Either the LMC has managed to totally destroy most of its older clusters (except for the handful of very ancient ones that may lay at a great distance from the LMC center), or the star formation rate in the LMC ~ 10 Gyr ago really was depressed relative to more recent times, as other evidence discussed below suggests. In this regard, absence of any really old NGC 2257-type objects in the SMC may be just compatible with the relative absolute magnitude differences between the two Clouds. On the other hand, the ~ 10 Gyr age of NGC 121 (Stryker et al. 1985) opens up not only questions about the age of RR Lyraes, which this cluster contains, but also on whether the SMC necessarily possesses any NGC 2257-aged stars at all.

There is obviously a great deal more work to be done in filling in the details of the Cloud age-abundance relations. It is still not known what the true scatter is about them, i.e. how big the abundance spread is at a given age. Another question of interest concerns how one can optimize integrated light measurements, such as the promising scheme extensively pursued by Searle and collaborators (e.g. Searle et al. 1980), to best resolve age-metallicity effects. This question is especially important with regard to the understanding of more distant extragalactic systems.

Unfortunately, progress along these lines may be slow until several current issues are cleared up. First, reliable abundance work for most Cloud clusters is desperately needed. There always seem to be nagging doubts about [Fe/H] measures from CM diagrams, but very little effort has yet been expended in obtaining the giant branch spectrophotometry with which one would like to check such values. Also, it would be nice if we knew how far away the Clouds were, as this again couples into the age. Although nobody originally set out looking for distances, virtually every group involved in cluster main sequence work has come to the conclusion that the data are far better fit by a "short-scale" LMC modulus of ~ 18.2 mag, rather than the canonical long-scale Cepheid modulus of ~ 18.7 mag. Schommer (1986) has shown that attempts to resolve this discrepancy by invoking convective overshoot effects (e.g. Chiosi 1986) do not work, so the problem remains with us.

2.2 The Star-Formation Rate

Have star bursts been of importance over the lifetimes of the Clouds, or has the star formation rate been more nearly constant? It is probably still too premature to extract an answer to this question from Figure 1, but there are many other lines of evidence pointing toward the conclusion that star formation in at least the LMC was greater $\sim$ 3-5 Gyr ago than in the more distant past. Quantifying the degree of this enhancement, however, has proven to be a somewhat elusive task.

Several observations which at first glance seem to support the notion of a strong burst turn out on closer examination to be quite compatible with a constant formation rate. For instance, a steady birthrate coupled with cluster luminosity fading and possible tidal disruption effects can adequately match the observed LMC cluster age distribution (cf. Mould and Aaronson 1982, Elson and Fall 1985b). Similarly, the number of Cloud carbon stars, which on average we expect to be $\sim$ 3-5 Gyr old, is not really so large in an absolute sense. Mould (1986) points out that with a reasonable choice of IMF, if every star between 1 and 2 $M_\odot$ becomes a C star then the SFR need only be $\sim$ 1/6 of its present day rate. Another way to make this point is to note that the total contribution of carbon stars to the bolometric light of the Cloud population older than $\sim$ 1 Gyr is only 8%, compared with the mean value of $\sim$ 25% that is found in the "pure" intermediate-age systems represented by the clusters; the peak in the C star luminosity function at $M_{bol} \sim$ -5 is also naturally produced by a constant birthrate (see Aaronson 1986).

On the other hand, certain results are not so easily discounted. For instance, Reid and Mould (1985) found that the total AGB luminosity function in several different LMC patches could be best fit by a star formation rate truncated $\sim$ 4 Gyr ago. The major uncertainties here are modeling of of the AGB evolution (which however now seems well under control because of the newer cluster studies), and incompleteness at the faintest magnitude bins. (Reid and Mould also find evidence for a more recent burst $\sim 10^8$ year ago, as has also been argued for by Frogel and Richer 1983, among others.) A similar apparent dearth of older stars, as indicated by the veritable absence of any developed subgiant branch, was found by Hardy et al. (1984) in their massive study of a northwest section of the LMC bar.

Perhaps the most convincing evidence for an SFR depressed in the distant past comes from the break in the LMC luminosity function at $M_V \sim$ +3 first observed by Butcher (1977). As discussed by Stryker (1984), quantitative modeling of this feature strongly suggests several-fold enhancement of the star formation $\sim$ 3-5 Gyr ago. The main caveat is again incompleteness at the faintest level, so the Hubble Space Telescope's ability to reach $M_V \sim$ +7 in the LMC is obviously going to provide the critical data needed here.

3.   OTHER LOCAL GROUP dI AND dE SYSTEMS

Given that even the next nearest dI, NGC 6822, is ~ 5 mag in distance modulus beyond the Clouds, it is not surprising that very little detail about the older population has been discerned in these systems.   Recently, an economical method for exploring the AGB in resolvable objects has been developed by the author, and also independently by H. Richer.   The technique involves direct imaging through two intermediate-band filters that monitor CN and TiO absorption at ~ 8000Å, which when combined with a broad band color for measuring temperature, enables straightforward identification of carbon and M-type stars.   Spectroscopy has confirmed the high reliability of this method, and the first carbon star spectra have now been obtained in a number of Local Group galaxies, including NGC 6822, IC 1613, WLM, NGC 205, M31, and M33 (see Mould and Aaronson 1986).   Richer and collaborators have ventured beyond the Local Group and identified a number of C star candidates in the Sculptor systems NGC 55 and NGC 300 (e.g. Richer et al. 1985).

Various efforts are at present underway to completely map the AGB surface distribution and luminosity function in a number of Local Group objects.   When these programs reach fruition, we should have considerably greater insight into the star formation history in galaxies of different morphological type, as well as new important input for models of AGB evolution.   An added bonus may come in regard to the distance scale, if it can be established that the C star luminosity function is universal.   We should emphasize that study of the AGB is not just important from the standpoint of understanding the stellar content in nearer galaxies, but also for modeling evolution at high redshift, because the AGB is often a substantial if not dominant component in composite stellar systems.   For example, in the passively evolving models of Renzini and Buzzoni (1986) having solar abundance and a Salpeter IMF, the AGB contribution to the bolometric luminosity rises from ~ 20% in a system 10 Gyr old to   50% in one 0.1 Gyr old. The author thus fervently hopes that a place for the intermediate band filters described above can be found in the second generation WF/PC for the HST that is now under construction.   Study of AGB carbon and M stars in galaxies as distant as the Virgo cluster should be well within reach.

In the meantime, we content ourselves with a brief description of the preliminary AGB survey carried out by Cook et al.   (1986) in selected 1.5' x 2.5' patches in the dI systems IC1613, WLM, and NGC 6822, as well as in M31 and M33.   The principal result obtained was the strong dependence of C/M ratio on parent galaxy absolute magnitude, presumably reflecting the roughly monotonic relation between the latter and average metal abundance.   Such a trend is consistent with the change in C/M ratio seen as we move from the Galactic bulge (where C stars are exceedingly rare) through the disk and onto the Magellanic Clouds; Cook et al. in fact demonstrated that the Clouds fit smoothly onto the (C/M, $\overline{M_V}$) relation defined by the five systems they examined.

Abundance can drive a change in C/M ratio in two ways. First, lowering the metals shifts the effective temperature of the giant branch blueward in the HR diagram and at the same time depresses the TiO bandstrength at fixed T_e, making it "harder" to produce M stars. Second, lowering the metals means that less C need be mixed to the surface in order to drive [C] > [O], and may furthermore increase the dredge-up efficiency, making it "easier" to produce C stars. Cook et al. (1986) have argued that, at least for systems having luminosities and abundances at or lower than the level of the Clouds, the former effect dominates. For instance, the ratio of the total C star numbers in the LMC and SMC is in good accord with their absolute magnitudes. We will take this point up farther below when examining the statistics of carbon stars in the dwarf spheroidals.

It is important to stress, however, that metal abundance is not the whole story. Though luminous AGB carbon stars are commonly found in the redder clusters in the Clouds as well as the general field, they are completely absent in galactic globulars. Hence, the numerous C stars also now identified in other dI systems is an indication of an underlying rich, intermediate-age population. The upshot of all this is that the AGB population in the Magellanic Clouds would appear to be perfectly consistent with their global properties, and that therefore star formation in the Clouds has probably not been unduly perturbed by tidal effects of the Milky Way. To firm up this conclusion we must await more complete AGB surveys in other late-type systems.

With regard the dE's, there are four within the Local Group --NGC 147, 185, 205, and M32 -- that are amenable to modern study in sights of good seeing. Deep CM diagrams have been obtained already for patches in the outer parts of NGC 147 and 205 by Mould et al. (1983, 1984). In both objects these authors find a broad giant branch suggestive of a real dispersion in abundance. The absence of a luminous AGB leads Mould et al. to conclude that the contribution of an intermediate age component can be no more than 10% in the perimeter of either dwarf. One should contrast this result, however, with the well-known presence of young stars in the nuclear regions of NGC 185 and 205. The AGB carbon stars found by Richer et al. 1984 in the latter system are an indication that residual star formation has probably proceeded over its entire lifetime. The mean abundances estimated by Mould et al. for NGC 147 and 205 are in general accord with the mass-metallicity relation inferred for brighter ellipticals, though one should again be careful about possible radial gradient effects in making such comparisons.

4.   THE DWARF SPHEROIDALS

Seven are known to surround the Milky Way, and three more have been found by van den Bergh (1972) around M31. One suspects that these are the single most common type of galaxy in the Universe. In the Local Group we have little trouble confusing these spheroidals with the more luminous dE types already discussed, but of course, the degree to which this distinction is arbitrary remains something of a topic for lively debate. In contrast to the older view that dSph, dE, and gE systems form a one parameter family, considerable evidence has accumulated to suggest that early-type systems come in two separate varieties, as discussed by Wirth and Gallagher (1984). On the one hand, we have the high surface brightness ellipticals, whose high mass end is defined by the gE's and whose low mass tail is represented by systems such as M32. Then there are the low surface brightness ellipticals, bracketed by objects such as NGC 205 at one extreme and by the lowest luminosity dSph systems at the other. Furthermore, in their stellar content, in their structural parameters, and (perhaps most significantly), in their mass-to-light ratios, dSph and low luminosity dI galaxies share much in common. Is this just a coincidence, or a sign of some deep-seated common heredity? Before addressing the question further, we summarize present-day observations of the spheroidals.

4.1 Some Basic Properties

In Table 1 we list absolute magnitudes, distances, core and tidal radii, central surface brightnesses, and horizontal branch types, along with similar parameters for ω Cen. The various values have been adopted from Zinn (1985), Aaronson and Mould (1985 b), Kormendy (1986), and Hodge (1971), updating some of the quantities from more recent work of Olszewski and Aaronson (1985, Ursa Minor), Saha et al. (1986, Carina), and Armandroff and Da Costa (1986, Sculptor).

Table 1

Some Dwarf Spheroidal Properties

Name	M_V (mag)	d (kpc)	r_c (kpc)	r_t/r_c	$\mu_0(V)$ (mag/$\square$")	HB type
Fornax	-12.6	140	.5	6	23.3	red
Leo I	-11.4	220	.3	3	23.5	red?
Sculptor	-11.1	80	.2	6	23.9	red
Leo II	-10.2	220	.2	4	23.9	red
Carina	- 9.4	100	.2	3	24.9	red
Draco	- 8.5	75	.15	3	25.4	red
Ursa Minor	- 8.5	65	.15	6	26.1	blue!

In their absolute magnitudes, the spheroidals overlap with galactic globulars; Draco and Ursa Minor are the lowest luminosity galaxies known, though it remains uncertain as to whether or not they actually define the lower limit that luminous galaxian masses can attain. Unfortunately, most of the spheroidal magnitudes must still be regarded with some degree of skepticism, seemingly changing with every new study that appears. The superficial resemblance in stellar content with Galactic globulars was initially made in the landmark paper on Draco by Baade and Swope (1961), although peculiarities with the variable stars and in the morphology of the giant branch were already hinting perhaps at the interesting things to come. Draco's red horizontal branch has proven to be the rule, rather than the exception, for both spheroidals and outer halo globulars.

The central surface brightnesses of the spheroidals are remarkably small, not only making some of these objects exceedingly difficult to recognize, but also begging the question of how star formation ever proceeded in such a low density environment. There is a general trend of surface brightness with absolute magnitude counter to that seen for luminous E-types. Kormendy (1985) has examined in detail this and other core parameter relations, finding in all cases large discontinuities with the bright elliptical loci.

4.2   The Carbon Stars

Carbon stars have now been identified in all seven halo spheroidals; some of their properties are summarized in Table 2. The census of bright C stars listed in column 2 is now reasonably complete in all these objects, thanks largely to the work of Azzopardi et al. (1985, 1986b).

Table 2
Carbon Stars in Dwarf Spheroidals

Name	N_C[a]	$N_{C,L}$[b]	M_{bol}^{max}[c] (mag)	Age[d] (Gyr)	%IA[e]
Fornax	~ 70	-3.19	-5.3	3	25
Leo I	19	-3.28	-4.5	9	20
Sculptor	8	-3.54	-4.3	10	5
Leo II	7	-3.23	-4.4	10	15
Carina	9	-2.81	-4.8	6	70
Draco	4	-2.80	-3.5	16	--
Ursa Minor	1	-3.40	-3.5	16	--
LMC	11,000	-3.40			
SMC	2,900	-3.22			

[a] Number of known C stars.
[b] $N_{C,L} \equiv \log N_C + M_V/2.5$.
[c] Bolometric magnitude of the brightest star (not a C star in the case of Ursa Minor).
[d] Age of most recent epoch of star formation based on M_{bol}^{max}.
[e] Percentage of population that is intermediate age based on fractional carbon star light.

However, there is still some question about the total number of
faint C stars.   For instance, included among the Draco total is the
marginal case noted by Azzopardi et al. (1986b), as a recent spectrum
obtained by the author and E. Olszewski does show strong C_2 bands, and
a preliminary radial velocity confirms membership.   Curiously, this
object was also noted as a possible C star on the Draco grism plate
obtained much earlier by the author (Aaronson et al. 1982), though a
confirming spectrum was not secured at that time because of the high
proper motion, apparently in error, listed by Stetson (1980).   The star
in question is No. 578 from Baade and Swope (1961), which at V = 18.25
(M_V = -1.1) makes it the faintest C star by far now known in a dwarf
spheroidal, placing it well below the tip of the giant branch, and
raising the question of how many such stars might be present but beyond
the prism plate limits in the more distant systems.   Note also that for
consistency the CH star known to be a member of Ursa Minor has not
been included in the total there, since both the Swan C_2 and near-red
CN bands are too weak for objects such as this to have been spotted
with either of these two most popular prism methods.   (It has often
been stated that the CN and C_2 band approaches are complementary in
that the former best selects red and/or reddened carbon stars, while
the latter method best finds blue ones.   However, the CN and C_2 band
strengths appear to be rather correlated, and as far as the author is
aware, there is yet to be a single well documented case where
comparable plate material exists of a star selected by one technique
being missed by the other -- but see Azzorpardi et al. 1986a.)

Additional information given in Table 2 includes the fractional
carbon star light (column 3), and the bolometric magnitude of the
brightest member star (column 4), being in all cases except Ursa Minor
a carbon star (see Aaronson and Mould 1986).   It can be seen that
except for Draco and Ursa Minor, the dwarf spheroidals all have
extended giant branches, terminating well above the first giant branch
tip, which by analogy with clusters in the Magellanic Clouds is prima
facie evidence for the presence of an intermediate-age population.   The
theory of the AGB (e.g. Aaronson and Mould 1985a) can be used to turn
these luminosities into a gauge of the most recent period of active
star formation, with the results given in column 5.   The ages there are
in good agreement with the results obtained from the deep
color-magnitude diagrams discussed below, though it would be useful in
this regard to have a more complete set of IR photometry, especially
for the C stars in Leo I.   The rough correlation between these ages and
the parent system absolute magnitudes suggests that the more massive
dwarfs were able to retain gas the longest to form successive
generations of stars.

By comparing the fractional carbon star light in column 3 of Table
2 with that in pure-age Cloud clusters, it is possible to derive a
measure of the population percentage that is of intermediate age, i.e.
younger than ~ 10 Gyr; this is given in the last column of Table 2.   It
should be emphasized that these estimates are crude at best and could
easily be off a factor of two.   Nevertheless, the results suggest that
in most of the dwarf spheroidals the majority of the stars were

produced early on; the one glaring exception is Carina, which we return to later. We note that the discovery of a luminous AGB carbon star in the Andromeda II system (Aaronson et al. 1985) implies that such stars and the intermediate-age population that accompanies them are likely to be a common feature of most spheroidal galaxies.

In Figure 2 we have plotted the fractional carbon star light from Table 2 against mean abundance (see below). Excluding the Milky Way, there is no pronounced trend with [Fe/H]. Put another way, the total number of carbon stars in these systems scales roughly as the absolute

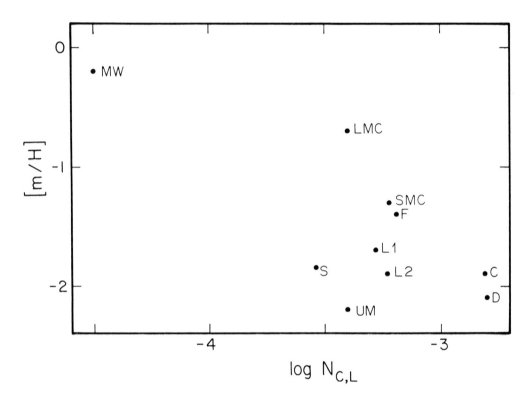

Figure 2. Fractional C star number ($N_{C,L} = \log N_C + M_V/2.5$) plotted against mean abundance for the spheroidals, the Clouds, and the Galaxy. For the latter three systems, the metallicity about 3 Gyr ago was adopted (see Figure 1), when the "typical" C star may have formed. The Milky Way value is based on the solar neighborhood (Richer and Westerlund 1983).

magnitude, a point already made in regard to the dI results discussed earlier in Section III. We note that these conclusions are opposite to those reached by Richer and Westerland (1983), who argued from preliminary statistics for a strong relation between carbon star number and abundance. However, we believe there may be a perfectly good theoretical justification for the lack of any trend in Figure 2. In particular, model calculations from Iben (1983) indicate that for a stellar core mass $M = 0.7M_\odot$ and $[Fe/H] = -1$, only one dredge-up episode is required to drive $[C] > [O]$ and make a carbon star. In other words, any further lowering of the abundance below $[Fe/H] \sim -1$ should not necessarily be expected to make C star production any "easier". A similar comment applies to observations of stars in Draco and Ursa Minor that possibly indicate enhanced carbon abundance (see Kraft 1986), unless of course primordial carbon stars are involved!

When one plots up the bolometric luminosities of the carbon stars that have good IR magnitudes, an interesting gap is present at $M_{bol} \sim -3.6$. This is roughly the magnitude below which dredge-up theory presently seems incapable of producing carbon stars. However, lower luminosity C stars have been found in Draco, Ursa Minor, Carina, and Sculptor; where do they come from? The most likely place is binaries, and indeed there is direct evidence from radial velocity measures that several of these stars are binary members (Aaronson and Olszewski 1986a), as are apparently the majority of CH stars in the Galactic halo (McClure 1984). We might expect such low luminosity stars to be present in the more distant spheroidals, and again a complete IR survey of the known C stars would be of some interest for checking this.

The binary hypothesis might also explain the general absence of such stars in globular clusters, where the tidal field probably disrupts such systems early on. An alternative explanation put forward by Kraft (1986) that the absence of cluster C stars is simply a statistical effect is not, I believe, entirely tenable. If in a dwarf spheroidal with $M_V \sim -8.5$ we find on average 2.5 low luminosity C stars (Table 2), then the median globular magnitude of ~ -7.3 (Harris and Racine 1979) suggests that we ought to find typically one C star in every four out of five globulars, assuming again the independence in carbon star production from $[Fe/H]$ argued for earlier. Given the hundreds of spectra collected in dozens of globulars, it seems surprising that ω Cen is the only cluster in which stars with strong C_2 bands (as opposed to merely CH stars) have been located. Here only ~ 5 such stars have been found (Cohen and Bell 1986), even after an extensive grism search (Bond 1975), rather than the 15 or so that might be expected.

4.3 Abundances and Ages

Quite a bit of work has been done on spheroidal abundances over the last few years using a variety of estimators, including the slope of the giant branch; $(B-V)_{0,g}$; IR photometry; width of the subgiant branch; and single star spectrophotometry. The last of these is what people generally attach highest weight to and is now available for all the halo systems except Carina and Fornax, albeit in some cases for only a handful of stars. In the first part of Table 3 are listed the author's own appraisal of where the mean abundances and their associated dispersions now stand, incorporating the most recent spectrophotometric results on Sculptor (Armandroff and Da Costa 1986),

Table 3

Spheriodal Abundances and Ages

Name	<[m/H]>	Metal Range	Age[a] Range (Gyr)
Fornax	-1.4	1.4	3 - "old"
Leo I	-1.7:	?	---
Sculptor	-1.85	.6	13 - "old"
Leo II	-1.9	?	---
Carina	-1.9	?	7 - "old"
Draco	-2.1	.6:	"old?"
Ursa Minor	-2.2	0[b]	"old"

[a] Blue stragglers also present in Ursa Minor, Draco, Carina, and Sculptor.

[b] Excluding one very metal poor star.

Ursa Minor (Suntzeff et al. 1986b), and the Leo dwarfs (Suntzeff et al. 1986a), with the older material summarized by Zinn (1985).

The newer data strengthen the result known for some time now that the relation between mean abundance and absolute magnitude is a strikingly good one, compared for instance with the large scatter seen in the similar relation that has been inferred for brighter ellipticals. The effect is obviously a fundamental one for models of spheroidal evolutionary history. However, while it has often been argued that the natural extension of the (M_V, [Fe/H]) dSph relation onto that for brighter ellipticals is an indication that the former are simply the low mass tail of the latter, both spiral and dI galaxies also exhibit an abundance-magnitude effect which can just as easily be shown to tie onto the spheroidal relation, as demonstrated in Figure 2 of Aaronson (1986).

The existence of a real abundance spread seems reasonably well established now in three of the spheroidals (Buonanno et al. 1985, Armandroff and Da Costa 1986, Carney and Seitzer 1986), and there is a hint that this spread is also related to absolute magnitude, as might be expected. The reality of the Draco abundance dispersion has always been something of a controversial issue, with Bell (1985) most recently questioning its existence. Nevertheless, the generally good agreement among recent observations for stars that have overlapping data combined with the subgiant branch width measured by Carney and Seitzer (1986) lend credence to the presence of a real spread. The situation in Ursa Minor remains curious, with the Suntzeff et al. data indicating no dispersion above the observational errors, except for one very metal poor star having [Fe/H] ~ -3.5.

Turning to the ages, it is interesting to first note that in the 1979 IAU Colloquium, Scientific Research with the Space Telescope, one can find several times mention of the importance this instrument will have for reaching the main-sequence turnoffs in the spheroidals and age-dating them directly. In the meantime, the main-sequence has been reached now in four (and probably five) of these systems using CCD's from the ground. This was first accomplished for Carina by Mould and Aaronson (1983), who found a turnoff age of only 7 Gyr. Perhaps the most startling result of this effort, though, was the apparent absence in the Carina luminosity function of a substantially older component, leading Mould and Aaronson to propose that the bulk of the stellar content in Carina was of intermediate age (see also Table 2 here). The rather complete survey for and discovery of RR Lyraes in Carina by Saha et al. (1986) now indicates that an older population is present, though the conclusion that this is a minority component still stands. Saha et al. found 53 RR Lyraes compared with the 133 discovered by Baade and Swope (1961) in Draco. If the production rate of RR Lyraes in these two systems is similar, their relative luminosities implies that only ~ 15% of the Carina population is ancient.

The second spheroidal to have its main sequence reached was Sculptor, where Da Costa (1984) found a turnoff age ~ 2-3 Gyr younger than for galactic globulars, with a probable age spread of a comparable size. Ursa Minor was next, which Olszewski and Aaronson (1985) found to be excellently fit by M92. It seemed that the predominance of red horizontal branches among the spheroidals could be largely understood as an age effect, rather than one of the other numerous possibilities, consistent with the slow collapse picture of the outer halo put forward by Searle and Zinn (1978). Only Ursa Minor, with its old stellar population and blue horizontal branch, appeared to be living up to the idea of a canonical Population II object. Results for Draco were thus awaited with some anticipation. While similar to Ursa Minor in total luminosity, the horizontal branch morphology and the abundance spread both pointed again toward age effects as the culprit. Two deep CM diagrams are now available for Draco, from Stetson et al. (1985) and from Carney and Seitzer (1986). Unfortunately, the expected age difference has been somewhat difficult to discern in these results. It may be that we are up against the limits of ground based photometry and age resolutions of ~ 2 Gyr are simply too much to ask for; we must await Space Telescope for the answer here. In the meantime, it is worth noting that blue stragglers have been found in the four spheroidals discussed so far, so that a minority younger component may in fact still be present in all of these systems.

Finally, we mention the Buonanno et al. (1985) data on Fornax. CM diagrams for four of the Fornax globulars obtained by these authors show them to have standard globular cluster-like morphology, with the presence, in particular, of blue horizontal branch stars, probably indicative of very old ages. On the other hand, in the field of Fornax, Buonanno et al. find a group of stars located in a clump below the horizontal branch, which they interpret to be the top of the main-sequence. The corresponding age is ~ 3 Gyr, precisely what one would expect for the progenitors of the luminous carbon stars (see Table 2). Hence, it appears that for most of its lifetime Fornax has been manufacturing stars.

4.4   The Variables

RR Lyraes are known to be present in all the spheroidals except for Fornax and Leo I, where adequate searches have yet to be made. In Draco, Leo II, and Carina the mean periods $\langle P_{ab} \rangle$ lie intermediate between the two Oosterhoff classes. Anomalous Cepheids have also been found in all the spheroidals (excepting Carina); the first such star to be discovered in Fornax was recently reported by Light et al. (1986). The underlying cause of the various unusual properties of the spheroidal variables is doubtlessly traceable to the spread in ages and abundances discussed above.

The first type II Cepheid to be found in a spheroidal was also reported by Light et al. (1986). These stars are thought to evolve from a blue (and presumably old) horizontal branch. It is therefore somewhat gratifying that the type II Cepheid appears to be a member of one of the aforementioned Fornax globulars.

The presence of spheroidal blue stragglers and their ratio with respect to the anomalous Cepheids can be considered consistent with the long held notion of a causal connection between these two types of objects. The question of whether such stars are young or originate from much older binaries is still open, though. Light et al. have suggested that the Fornax anomalous Cepheid may, for a single star, be too metal rich to cross into the instability strip (e.g. Hirschfeld 1980). It is interesting that Butler et al. (1982) measured a mean abundance of [Fe/H] = -1.3 for several of the anomalous Cepheids in the SMC having a mean pulsation mass near 1.3 M_O. This abundance and mass is comparable to what might be expected for the Fornax anomalous Cepheid, so possibly all these stars are binaries.

On the other hand, as Smith and Stryker (1986) point out, the young star hypothesis provides a natural explanation of why no anomalous Cepheids are found in the LMC, given the Cloud enrichment histories depicted in Figure 1. If older binaries are the source of these stars, it is difficult to see why only the SMC contains them since both Clouds appear to have had roughly the same [Fe/H] about 10 Gyr ago. However, in more recent times, the Large Cloud would have become too metal rich to produce anomalous Cepheids -- but not the Small Cloud. While this all sounds appealing, one must still contend with the presence of an anomalous Cepheid in the galactic globular NGC 5466. It would be interesting to test for the binary nature of the latter, which is within the limits of present technology.

4.5   Mass-to-Light Ratios

In the last few years, a number of efforts have been made to measure directly the gravitating mass of various spheroidals and determine whether, like their larger-size cousins, these objects contain non-luminous matter. This question is not only of obvious importance for models of the origin and evolution of the dwarfs, but also for the nature of the dark material itself. In particular, phase space density arguments probably rule out the idea that the dark stuff is hot, i.e. in the form of massive neutrinos (see Madsen and Epstein 1985, Tremaine and Gunn 1979).

Progress in this area is somewhat hampered by the need to measure velocities of a few km s^{-1} accuracy for stars having V magnitudes > 17, a generally time consuming process even with the few instrumental set-ups with which such observations can be made. A summary up to the summer of 1985 of what has been something of a controversial situation is given in the fine review article by Kormendy (1986). Since then some interesting developments occurred. First, Armandroff and Da Costa (1986) have determined velocities for a large number of K giants in Sculptor and find a dispersion similar to that reported for carbon stars by Seitzer and Frogel (1985). Second, Aaronson and Olszewski (1986b) have obtained velocities for some of the Fornax globulars and again see a dispersion roughly the same as what Seitzer and Frogel found from carbon stars. Seitzer and Frogel had suggested that the similarity of the dispersion they measured from C stars in three dwarf spheroidals was perhaps related to the suspect nature of using carbon stars for any such measurement, but this appears not to have been the case.

Table 4 summarizes the present status of the velocity dispersion observations in five spheroidals. The dispersion errors listed are based on root N statistics only (though the dispersions themselves generally have been adjusted for the quoted velocity errors) and are usually a dominant source of uncertainty in any mass estimate. The second column of Table 4 lists the quantity $[(Lr_t)/(Lr_t)_{Fornax}]^{1/2}$. It is well to remember that under the most naive structural assumptions, a constant M/L leads to these factors by which we might expect the measured dispersion to decrease relative to Fornax. A comparison of columns 2 and 3 indicates, however, that something very interesting is going on. Total M/L values calculated from King models are given in column 4, and it can be seen there is a general increase of this quantity with decreasing luminosity. Almost any method used to calculate M/L leads to the same conclusion. For instance, Kormendy's (1986) approach based on central volume densities gives, if anything, slightly bigger M/L values than in Table 4. The results for Draco and Ursa Minor may seem uncomfortably large. Kormendy (1986) argues that these systems would contain a dark matter component with a higher central mass density than is suspected in any other galaxy, and therefore suggests that something may be wrong. However, since the systems in question are the two lowest luminosity galaxies known, we should perhaps not be too surprised by other extreme properties they might exhibit.

It is important to emphasize that Draco and Ursa Minor are still the only dwarfs where multiple epoch observations for the same stars

Table 4

Spheroidal Velocity Dispersions

Name	$\left[\dfrac{L/r_t}{(L/r_t)_{Fornax}}\right]^{1/2}$	$\sigma \pm \sigma/\sqrt{2N}$ (km s^{-1})	M/L$_V$	Sample	Ref[a]
Fornax	1	7.5 $\pm$ 3		3 clusters	1
		6.5 $\pm$ 2	2	5 C stars	2
Sculptor	.8	6 $\pm$ 2.5		3 C stars	2
		6.5 $\pm$ 1	5	16 K stars	3
Carina	.5	5.5 $\pm$ 1.5		6 C stars	2
		6 $\pm$ 2	8	5 C stars	4
Draco	.4	9 $\pm$ 2[b]	40	9 (2C,7K)	5
Ursa Minor	.3	11 $\pm$ 2[b]	100	7 K stars	5

[a] References: 1 - Aaronson and Olszewski 1986b; 2 - Seitzer and Frogel 1985; 3 - Armandroff and Da Costa 1986; 4 - Cook et al. 1983; 5 - Aaronson and Olszewski 1986a.

[b] Multiple epoch observations.

are available. Since the report by Aaronson and Olszewski (1986a), five additional radial velocities have been secured, including two more for the two extreme-velocity members of Ursa Minor, and one more for one of the extreme-velocity Draco stars, with velocity stability having been maintained in all cases. These and earlier multiple-epoch results rule out the influence of atmospheric motions. The Monte Carlo simulations reported by Aaronson and Olszewski (1986a) also imply that if the binary characteristics in the spheroidals are at all like those in the Galactic disk, then unless these workers have been unusually unlucky, duplicity effects should also not be of concern.

In Figure 3 we have plotted M/L as a function of absolute magnitude for the spheroidals in Table 4 and for the low luminosity dI systems mapped in H I by Sargent and Lo (1986). An overall trend is apparent, and with another 1-2 magnitudes of fading, the least luminous dI's would develop M/L ratios comparable with those seen in Draco and Ursa Minor. As Kormendy (1986) notes, the M/L values for Fornax, Carina, and Sculptor (the dSph systems which appear to lie below the dI systems in Figure 3) should probably be considered as lower limits:

although small, the central stellar mass densities in these objects are nonetheless greater than those of "plausible" dark halos.

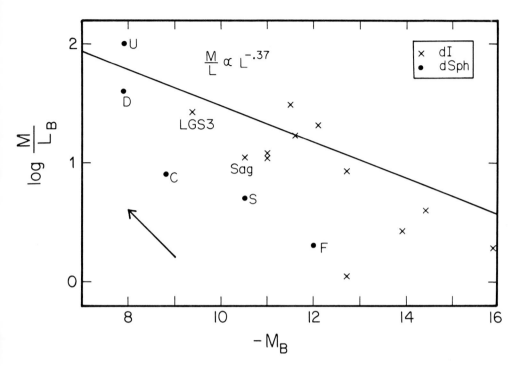

Figure 3. The mass-to-light ratio is shown as a function of absolute magnitude for the spheroidals in Table 4 (assuming B-V = 0.6 mag) and the irregulars from Sargent and Lo 1986. The arrow in the lower left represents one magnitude worth of fading. The line running diagonally across the diagram is the predicted relation from Dekel and Silk 1986.

4.6 Origins

Formation models for the spheroidals usually fall into two categories: disruptive and isolated. The disruptive models generally posit the spheroidals to be born from the debris of one or more tidal encounters between the Milky Way and the Magellanic Clouds. The positive aspects of this picture are that first, it explains the roughly co-planar distribution of the spheroidals on the sky; and second, the entire problem of how star formation proceeds in such low density conditions is swept, as it were, under the rug.

However, at this time it would seem that the disruptive scenarios have some major drawbacks. For one thing, they cannot account for what is now known about the stellar content of the spheroidals in anything but a very ad hoc fashion, and it is not clear how they could explain at all the high mass-to-light ratios, should those prove to be correct. For instance, it is difficult to see why the ([Fe/H], M_V) relation should be so good. As Zinn (1985) points out, one might have instead expected a tight ([Fe/H], age) relation, which is belied by a system like Carina. In fact, in most of the spheroidals it appears that star formation must have proceeded well after the separation event. The number and apparent very old age of the globulars in Fornax also make it hard to see how this system originated from the Clouds.

It is thus more tempting to conclude that the spheroidals developed in relative "isolation", though this word should be taken with a grain of salt, as it may well be that the formation of dSph systems hinges upon the presence of infalling gas around large-size seed galaxies, and that once formed tidal effects do play a role in the evolution of these objects. Now various authors have tried applying the simple model of chemical evolution to the spheroidals, for which

$$z = - y \ln(M_G/M) ,$$

where abundance z is a function of both the fractional gas mass M_G/M and the yield y, often but possibly incorrectly assumed to be a constant (e.g. Mould 1984). If the simple model is driven to completion, then the mean abundance <z> approaches the yield. The fact that the abundances in the spheroidals are so low is usually interpreted as an indication that these systems lost their gas mass early on, with an object like Draco having perhaps lost 99% of its total original mass.

Such a drastic process raises some concerns. Supernovae driven winds are the most popular method for depleting the gas (cf. Smith 1985, Vader 1986), but this cannot happen over too rapid a timescale, else the system is disrupted. However, with adiabatic gas removal, the product of the total mass times the characteristic radius MR stays constant (see Smith 1985), implying that a system like Draco began 100 times smaller and 10^6 times more dense, which seems a little fishy. Furthermore, Dekel and Silk (1986), argue that such a scenario cannot correctly reproduce the ([Fe/H], M_V) and ([Fe/H], surface brightness) relations unless a dark halo is present (which also alleviates the aforementioned MR problem).

Immersing spheroidals in a dark halo provides an obvious means by which the M/L values of these systems can vary. In particular, if in the smallest systems the star formation epoch was cut short by early gas removal, we might expect to see an increase in M/L with decreasing luminosity. Dekel and Silk (1986) use the observed (luminosity, surface brightness) relation to normalize the expected trend, and predict that M/L α $L^{-.37}$. This relation is drawn onto Figure 3, and the overall agreement seems rather impressive.

Of course, if one does not like dark matter, an alternative explanation for the results in Figure 3 involves changing the IMF. There appears to be accumulating evidence that the slope of the IMF varies with abundance in the sense needed, coming at the high mass end from observations of giant H II regions (Terlevich and Melnick 1983) and at the low mass end from globular cluster luminosity functions (McClure et al. 1986). With a judicious allowance for remnants, the high M/L values seen for Draco and Ursa Minor might just be accommodated by such a scheme.

Finally, we return to the question of whether dI's turn into dSph systems. It has become fashionable to address this issue by turning to the Virgo cluster, where large numbers of dE's are found, but the applicability of this sample to the Local Group spheroidals is possibly open to question. Morphologically there are certainly many cases of transitional-looking objects in Virgo (e.g. Sandage and Binggeli 1984), though a causal connection does not follow. On the other hand, inferred abundances, flattening parameters, the presence of nucleated dE's, and various other structural characteristics have led several authors (cf. Thuan 1985, van den Bergh 1986, Bothun et al. 1986) to argue against the dI → dE conversion, though the present author does not find all of these arguments entirely compelling (see Aaronson 1986). A related possibility is that, while there may be today no apparent dI progenitors of the dE's, the Virgo dI's that are seen will eventually turn into objects more like the nearby spheroidals (see Bothun et al. 1986).

Another approach to the problem is to try to determine the rotational properties of the spheroidals, many of which are rather elongated. A first step in this direction has been taken by Paltoglou and Freeman (1986), who have measured a value of $v/\sigma < 1$ for Fornax, which seems again to argue against the idea of a stripped irregular for the origin of this system. Even so, one does not yet have a good feel for what the v/σ values for low luminosity dI systems should be like. Indeed, in most of the small dI's studied by Sargent and Lo (1986), chaotic rather than rotational motion was detected.

Are there local examples of things which might be considered transitional between dI and dSph types? The smallest dI's are the best place to look for such objects, and we shall discuss the two such lowest luminosity objects known -- the Sagittarius dwarf ($M_B \sim -10.5$) and LGS 3 ($M_B \sim -9.4$). The first thing to notice about these systems is that their absolute magnitudes (though uncertain because of poor distances) lie in the same regime as the dSph's. The second point is that their M/L ratios are quite large, as noted in Figure 3. Recently, K. Cook and the author took a look at the red stellar content of SagDIG using the method described earlier, and the results are shown in Figure 4. After subtracting off galactic foreground contamination, the upper AGB of this system appears to be populated entirely by carbon stars, and in this respect it is very similar to the Fornax dwarf. However, LGS 3 is even more intriguing. Christian and Tully (1983) have pointed out that the brightest stars in this object are red, and

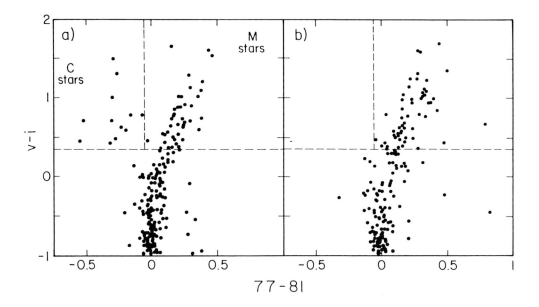

Figure 4. (a) Instrumental 77-81 color versus instrumental v-i color for the Sagittarius dwarf irregular. Regions in which C and M types fall are blocked out. (b) Same as (a) for a control region of roughly equal area, showing that virtually all the M stars in SagDIG are in the Galactic foreground.

that star formation in it has not occurred for the last 10^8 years, if not much longer. This system would be called a spheroidal, except for the fact that it contains H I, though not very much of it. Sargent and Lo (1986) find that M_{HI}/M_T is only 0.01 for LGS 3 (with values nearly this low being not that uncommon among the rest of their sample). Nevertheless, the gas content of LGS 3 is still far greater than the limits that have been set for the spheroidals.

Perhaps, however, the whole question of whether dI's turn into dE's is beside the point, as Dekel and Silk (1986) have emphasized in considering the spectrum of density fluctuations in models of biased galaxy formation. In their view, both types arise from one-sigma peaks, whereas more massive galaxies come from three-sigma fluctuations. The rarity of M32-type systems among the dE's then reflects the uncommoness with which stronger peaks give rise to low mass galaxies.

5. FUTURE WORK

It is hopefully obvious from the preceding discussion that there is still much to be done. For instance, present-day estimates of the structural parameters in the nearby spheroidals trace back almost entirely to the massive effort of Paul Hodge in the early sixties. As one can go much fainter today, it would be nice to have this work repeated with modern plates and analysis methods (i.e. we need another hero). The detailed mapping of the AGB that several groups are now pursuing in the Magellanic Clouds and other Local Group objects can be expected to provide us with a clearer window into the star formation history of these systems long ago. However, doubtlessly the really exciting results in this area will come with the Hubble Space Telescope. The deep color-magnitude diagrams and luminosity functions that this instrument will provide for us should enable a much better picture of the stellar age distribution in nearby dwarfs to be put together. Finally, we need far more work done on dwarf abundances and kinematics, fitting problems for the coming generation of large ground-based telescopes. One also suspects that there may be a considerably greater number of systems like Draco floating about out there, but finding them presents a challenge. A novel approach here, already tried once (though unsuccessfully) by van den Bergh and Lafontaine (1984), would be to look for such objects around very faint and distant carbon stars. In any event, we can rest assured that the dwarf galaxies within the Local Group will remain a fascinating laboratory in which to explore the nature of stellar populations for many years to come.

It is a pleasure to thank Ed Olszewski and Kem Cook for helpful discussion, and the latter for the use of unpublished data. Preparation of this review was partially supported with funds from NSF grant AST83-16629. The Mt. Wilson and Las Campanas Observatory, where part of the article was prepared, is also acknowledged for inviting the author to an enjoyable summer visit.

REFERENCES

Aaronson, M. (1986). In Star Forming Dwarf Galaxies and Related Objects, ed. D. Kunth and T. X. Thuan, in press.
Aaronson, M., Gordon, G., Mould, J., Olszewski, E., and Suntzeff, N. (1985). Ap. J. (Letters), **296**, L7.
Aaronson, M., Liebert, J., and Stocke, J. (1982). Ap. J., **254**, 507.
Aaronson, M., and Mould, J. (1985a). Ap. J., **288**, 551.
_____ (1985b). Ap. J., **290**, 191.
Aaronson, M., and Olszewski, E. W. (1986a). In IAU Symposium No. 117, Dark Matter in the Universe, ed. J. Kormendy and G. R. Knapp, in press. Dordrecht: Reidel.
_____ (1986b). A. J., in press.

Armandroff, T. E., and Da Costa, G. S. (1986). Poster paper presented at Conference on Stellar Populations, Space Telescope Science Institute, Baltimore.

Azzopardi, M., Dumoulin, B., Quebatte, J., and Rebeirot, E. (1986a). ESO Messinger, No. 43, p. 12.

Azzopardi, M., Lequeux, J., and Westerlund, B. E. (1985). Astr. Ap., 144, 388.

_____ (1986b). Astr. Ap., in press.

Baade, W., and Swope, H. H. (1961). A. J., 66, 300.

Bell, R. A. (1985). Publ. A.S.P., 97, 219.

Blanco, V. M., and McCarthy, M. F. (1983). A. J., 88, 1442.

Bond, H. E. (1975). Ap. J. (Letters), 202, L47.

Bothun, G. D., Mould, J. R., Caldwell, N., and MacGillivray, H. T. (1986). Preprint.

Bruck, M. T., Cannon, R. D., and Hawkins, M. R. S. (1985). M.N.R.A.S., 216, 165.

Buonanno, R., Corsi, C. E., Fusi Pecci, F., Hardy, E., and Zinn, R. (1985). Astr. Ap., 152, 65.

Butcher, H. R. (1977). Ap. J., 216, 372.

Butler, D., Demarque, P., and Smith, H. (1982). Ap. J., 257, 592.

Carney, B., and Seitzer, P. (1986). Preprint.

Chiosi, C. (1986). In Spectral Evolution of Galaxies, ed. C. Chiosi and A. Renzini, p. 237. Dordrecht: Reidel.

Christian, C. A., and Tully, R. B. (1983). A. J., 88, 934.

Cohen, J. G., and Bell, R. A. (1986). Ap. J., 305, 698.

Cook, K. H., Aaronson, M., and Norris, J. (1986). Ap. J., 305, 634.

Cook, K. H., Schechter, P., and Aaronson, M. (1983). Bull. A.A.S., 15, 907.

Da Costa, G. S. (1984). Ap. J., 285, 483.

Dekel, A., and Silk, J. (1986). Ap. J., 303, 39.

Elson, R. A. W., and Fall, S. M. (1985a). Publ. A.S.P., 97, 692.

_____ (1985b). Ap. J., 299, 211.

Freeman, K. C., Illingworth, G., and Oemler, A., Jr. (1983). Ap. J., 272, 488.

Frogel, J. A. (1984). Publ. A.S.P., 96, 856.

Frogel, J. A., and Richer, H. B. (1983). Ap. J., 275, 84.

Hardy, E., Buonnano, R., Corsi, C., Janes, K., and Schommer, R. (1984). Ap. J., 278, 592.

Hardy, E., and Durand, D. (1984). Ap. J., 279, 567.

Harris, W. E., and Racine, R. (1979). Ann. Rev. Astr. Ap., 17, 241.

Hesser, J. E., McClure, R. D., and Harris, W. E. (1984). In IAU Symposium No. 108, Structure and Evolution of the Magellanic Clouds, ed. S. van den Bergh and K. S. de Boer, p. 47. Dordrecht: Reidel.

Hirshfeld, A. W. (1980). Ap. J., 241, 111.

Hodge, P. W. (1971). Ann. Rev. Ast. Ap., 9, 35.

Hodge, P. W., and Wright, F. W. (1967). The Large Magellanic Cloud. Washington: Smithsonian Press.

Iben, I. (1983). Ap. J. (Letters), 275, L65.

Kormendy, J. (1985). Ap. J., 295, 73.

_____ (1986). In IAU Symposium No. 117, Dark Matter in the Universe, ed. J. Kormendy and G. R. Knapp, in press. Dordrecht

Reidel.
Kraft, R. (1986). In ESO Workshop on Production and Distribution of C, N, O Elements, ed. I. J. Danziger, F. Matteucci, and K. Kjar, p. 21. Garching: ESO.
Light, R. M., Armandroff, T. E., and Zinn, R. (1985). Bull. A.A.S., 17, 883.
Madsen, J., and Epstein, R. (1985). Phys. Rev. Lett., 54, 2720.
Mateo, M., Hodge, P. W., and Schommer, R. A. (1986). Preprint.
Mathewson, D. S., and Ford, V. L. (1984). In IAU Symposium No. 108, Structure and Evolution of the Magellanic Clouds, ed. S. van den Bergh and K. S. de Boer, p. 125. Dordrecht: Reidel.
McClure, R. D. (1984). Ap. J. (Letters), 280, L31.
McClure, R. D., et al. (1986). Preprint.
Mould, J. (1984). Publ. A.S.P., 96, 773.
_____ (1986). In Spectral Evolution of Galaxies, ed. C. Chiosi and A. Renzini, p. 133. Dordrecht: Reidel.
Mould, J., and Aaronson, M. (1982). Ap. J., 263, 629.
_____ (1983). Ap. J., 273, 530.
_____ (1986). Ap. J., 303, 10.
Mould, J., Kristian, J., and Da Costa, G. S. (1983). Ap. J., 270, 471.
_____ (1984). Ap. J., 278, 575.
Olszewski, E. W., and Aaronson, M. (1985). A. J., 90, 2221.
Olszewski, E. W., Harris, W. E., and Schommer, R. A. (1986). In preparation.
Pagel, B. E. J. (1981). In Structure and Evolution of Normal Galaxies, ed. S. M. Fall and D. Lynden-Bell, p. 211. Cambridge: Cambridge University Press.
Paltoglou, G., and Freeman, K. C. (1986). Poster paper presented at IAU Symposium No. 27, Structure and Dynamics of Elliptical Galaxies, Princeton.
Reid, N., and Mould, J. (1985). Ap. J., 299, 236.
Renzini, A., and Buzzoni, A. (1986). In Spectral Evolution of Galaxies, ed. C. Chiosi and A. Renzini, p. 195. Dordrecht: Reidel.
Richer, H. B., Crabtree, D., and Pritchet, C. (1984). Ap. J., 287, 138.
Richer, H. B., Pritchet, C., and Crabtree, D. (1985). Ap. J., 298, 240.
Richer, H. B., and Westerlund, B. E. (1983). Ap. J., 264, 114.
Saha, A., Monet, D. G., and Seitzer, P. (1986). Preprint.
Sandage, A., and Binggeli, B. (1984). A. J., 89, 919.
Sargent, W. L. W., and Lo, K.-Y. (1986). In Star Forming Dwarf Galaxies and Related Objects, ed. D. Kunth and T. X. Thuan, in press.
Schommer, R. A. (1986). In Galaxy Distances and Deviations from Uniform Hubble Flow, ed. B. F. Madore and R. B. Tully, in press. Dordrecht: Reidel.
Schommer, R. A., Olszewski, E. W., and Aaronson, M. (1986). In preparation.
Searle, L., Wilkinson, A., and Bagnuolo, W. (1980). Ap. J., 239, 803.
Searle, L., and Zinn, R. (1978). Ap. J., 225, 357.
Seitzer, P., and Frogel, J. A. (1985). A. J., 90, 1796.
Smith, G. H. (1985). Pub. A.S.P., 97, 1058.
Smith, H. A., and Stryker, L. L. (1986). Preprint.
Stetson, P. B. (1980). A. J., 85, 387.

Stetson, P. B., VandenBerg, D., and McClure, R. D. (1985). Pub. A.S.P., **97**, 908.

Stryker, L. L. (1984). <u>In</u> IAU Symposium No. 108, Structure and Evolution of the Magellanic Clouds, ed. S. van den Bergh and K. S. de Boer, p. 79. Dordrecht: Reidel.

Stryker, L. L., Da Costa, G. S., and Mould, J. R. (1985). Ap. J., **298**, 544.

Suntzeff, N. B., Aaronson, M., Olszewski, E. W., and Cook, K. (1986a). A. J., **91**, 1091.

Suntzeff, N. B., Olszewski, E. W., Kraft, R. P., Friel, E., Aaronson, M., and Cook, K. (1986b). In preparation.

Terlevich, R., and Melnick, J. (1983). ESO Preprint No. 264.

Thuan, T. X. (1985). Ap. J., **299**, 881.

Tremaine, S., and Gunn, J. E. (1979). Phys. Rev. Lett., **42**, 467.

Twarog, B. A. (1980). Ap. J., **242**, 242.

Vader, J. P. (1986). Ap. J., **305**, 669.

van den Bergh, S. (1972). Ap. J. (Letters), **171**, L31.

_____ (1986). A. J., **91**, 271.

van den Bergh, S., and Lafontaine, A. (1984). Pub. A.S.P., **96**, 869.

Wirth, A., and Gallagher, J. S. (1984). Ap. J., **282**, 85.

Zinn, R. (1985). Mem. Soc. Astron. Ital., **56**, 223.

DISCUSSION

LUPTON: Is the spread in color of the main sequence in your CM diagram for Ursa Minor consistent with being purely due to photometric errors?

AARONSON: Yes, basically.   This point was examined in detail in Olszewski and Aaronson 1985.   On the main sequence, the sigma obtained from the scatter of the data is about 50% larger than what is calculated from just the photon statistics of the stellar images.   This no doubt reflects the additional sources of error one has to contend with (crowding, undersampling, sky determination, etc.).   A real (but small) spread in abundance may also be a contributing factor.

LARSON: I was struck by a point that you didn't emphasize, namely that there seems to be a close inverse proportionality between metal abundance and mass-to-light ratio among the dwarf spheroidals.   If so, this is just what would be expected for a simple model of chemical evolution if the mass is mostly in low-mass stars or in the remnants of very massive stars that collapse to black holes.   The simple model may then work after all, with no need to postulate extensive mass loss.   The variation in both Z and M/L with luminosity would in this case result from a variation' in the IMF.

AARONSON: Yes, and I will try to stress this point more in the written version of the talk.

HARRIS: Concerning the M/L ratios for the dwarf spheroidals: an additional point of support comes from the old problem that the observed tidal radii for these systems were regarded to be too large for their present solar galactocentric distances, if their adopted M/L was near 1-2 (similar to globular clusters).   In a paper three years ago by Innanen, Webbink, and myself, we discussed this problem generally for the halo clusters and concluded (for the dwarf spheroidals particularly) that the answer was (a) their adopted M/L was too low, (b) their outer parts were not yet in tidal equilibrium with the Galaxy because of the very long orbital times; or (c) the star counts from which the tidal radii were derived are systematically wrong.   But if M/L is as high as you say, then the observed tidal radii could be as large as they are with no problem.

AARONSON: Yes, I agree.   It was my impression that M/L arguments based on tidal radii were viewed with great suspicion by most people because of the second point that you mention.

ZINNECKER: The percentages you gave for the fraction of intermediate age populations in the nearby dwarf spheroidals presumably refer to the (infrared) light.   What fraction of the mass in these galaxies do you estimate is of intermediate age?

AARONSON: In the simplest of worlds, it would be roughly the same, but the answer really depends on what the IMF is doing.

SMITH: Is there evidence for a massive dark halo in the LMC?

AARONSON: I think problems in handling the Magellanic Stream have limited attempts to deal with this question, but I believe the evidence continues to accumulate that other Magellanic systems do have such halos.

KING: Regarding a possible large mass for the LMC, its dynamical friction in the massive halo of the Milky Way would then become embarrassingly large.

AARONSON: Tremaine (1976, Ap. J., 203, 72) has looked into the effects of dynamical friction on the Clouds. While the Clouds may decay into our Galaxy in the near future, I am not sure that much constraint is really put on their past history.

VAN DEN BERGH: What constraints can be placed on the past mass loss of dwarf spheroidals from the fact that many of them are observed to be significantly flattened?

AARONSON: It was my impression that the flattening characteristics of the spheroidals were, within small number statistics, similar to ellipticals in general, so I am not sure one can really say anything here.

DA COSTA: Can you estimate effects of differing stellar populations (of say Carina vs Sculptor or Ursa Minor) on apparent correlation of M/L with L?

AARONSON: I suppose so, but one probably wants some good quality HST data to work with first.

BOND: Are all carbon stars necessarily representative of an intermediate-age population? I am thinking of Omega Centauri, which contains carbon stars but is not of intermediate age.

AARONSON: I would say that when luminous carbon stars are present, i.e. those having $M_{bol} < -4$, an intermediate age population is not far behind. I know of no cases, in either the spheroidals or the Cloud clusters, where deep CM diagrams have failed to show this to be true. However, with the low luminosity C stars ($M_{bol} > -3.5$), dredge-up theory has a problem. These stars may be produced in binaries, in which case all bets about the age are off.

GLOBULAR CLUSTERS IN LOCAL GROUP GALAXIES

Robert Zinn

Yale Astronomy Department, P.O. Box 6666,
New Haven, Connecticut   06511, U.S.A.

INTRODUCTION

One of the major goals of research on star clusters is to provide
sufficient data (e.g., ages, metallicities, and kinematics) to piece
together a comprehensive picture of a galaxy's chemical and dynamical
evolution. While this goal has yet to be attained for any galaxy, the
most progress has been made in the galaxies of the Local Group.

For the star clusters of the Milky Way, the classification
globular cluster, which was originally assigned on the basis of a
cluster's appearance on the sky, has come to mean very old cluster
regardless of appearance (witness, for example, the minuscule globular
cluster Pal 13). In other galaxies, however, clusters are still
classified on the basis of appearance, and the brightest and most
populous ones are called globular clusters regardless of their ages
(witness the $\sim 10^7$ year old globular clusters in the Magellanic Clouds).
This well entrenched tradition will be followed here even though it has
led to some confusion in the past. Our discussion starts with the
globular clusters in the Milky Way, which are all very old, and then
branches out to the globular clusters in the nearest and the most
prominent members of the Local Group. In nearly all of these galaxies,
the globular clusters span a considerable range in age. The discussion
emphasizes the characteristics of the cluster systems that reveal the
most information about the evolution of their parent galaxies and the
similarities and dissimilarities that exist among these cluster
systems.

THE MILKY WAY

A large body of data exists for the roughly 150 known globular
clusters in the Milky Way. For example, measurements of metallicity
and radial velocity exist for more than 100 clusters, and
color-magnitude diagrams have been constructed for more than 90. In
more than 20 clusters the main-sequence turnoff has been reached
enabling one to estimate the ages of the cluster. These data have
revealed several characteristics of the cluster system that have major
implications for the evolution of the Galaxy (see Freeman and Norris
1981; Hesser 1983; Carney 1984 for other reviews). While it is not

within our current capabilities to study every one of these features
(if they exist) in the other members of the Local Group, this may
change with the flight of the Hubble Space Telescope (HST).

The Halo and Disk Subsystems

The recent improvements in the data for many clusters have made
much more tractable the question of whether subpopulations exist within
the Galaxy's globular cluster system. Baade (1958) suggested long ago
that the most metal rich globular clusters constitute a disk system, in
contrast to the halo population of metal-poor clusters. Until
recently, the evidence in favor of two populations of clusters has not
been convincing. Now there is more evidence for a metal-rich disk
system from both the spatial distribution and the kinematics of the
clusters (Zinn 1985), and some of this evidence is reviewed here.

The metal-rich clusters occupy a much smaller volume of space near
the galactic center than do the metal-poor clusters, which are spread
throughout the vast halo. Consequently, the observed spatial
distribution of the metal-rich clusters is distorted more by the
interstellar obscuration near the galactic plane. To see if the
distributions of the metal-rich and metal-poor clusters are flattened
by different amounts towards the plane, as expected if the metal-rich
clusters constitute a disk system, it is essential to restrict the
sample of metal-poor clusters in order to obtain one that has been
influenced by the obscuration in much the same way as the sample of
metal-rich clusters. Since the vast majority, if not all, of the
metal-rich clusters are closer to the galactic center than the sun, it
seems reasonable to compare them with the metal-poor clusters that also
lie within the solar distance ($R_\odot$). To avoid any biases caused by the
uncertain distance modulii of the clusters, their distributions on the
sky are compared. In the top graph of Figure 1, $<\theta_1>$, the mean value
of θ_1 for each group of clusters, is plotted against the mean
metallicity of each group (on the metallicity scale of Zinn and West
1984 which is used throughout this paper). Following Frenk and White
(1982), θ_1 is the angle between the galactic equator and the great
circle on the sky that passes through a cluster and the galactic
center. $<\theta_1>$ is expected to equal 45 degrees if the clusters have a
spherical spatial distribution about the galactic center. Figure 1
shows that $<\theta_1>$ is less than 45° for the clusters that have
[Fe/H] > -0.8, which indicates that they are flattened towards the
galactic plane. In contrast, the more metal-poor clusters have values
of $<\theta_1>$ greater than 45°. Since the distribution of the metal-poor
clusters beyond $R_\odot$ is roughly spherical, this effect suggests that
several clusters having small values of θ_1 have been excluded from the
sample because they are hidden by the interstellar dust clouds near the
galactic plane. There is every reason to believe that several
metal-rich clusters have also been hidden, and therefore, that the
value of $<\theta_1>$ plotted in Figure 1 for the metal-rich group has been
biased upward.

The other graphs in Figure 1 show the results of calculating, by the method of Frenk and White (1980), the mean rotational velocity, V_{rot}, and the line of sight velocity dispersion, σ_{los}, for each of the groups of clusters in the top graph, minus the few clusters for which no measurements of radial velocity and distance exist. The quantity V_{rot}/σ_{los} is a useful diagnostic of rotational flattening (see Binney 1978), which is expected to be greater than one in systems flattened by rotation and less than one in slowly rotating halo systems. The data in Figure 1 show that the clusters more metal rich than [Fe/H] = -0.8 have the kinematics of a flattened system, in good agreement with the data plotted in the top graph. From Figure 1 and the other data presented by Zinn (1985), it appears that flattening, V_{rot} and V_{rot}/σ_{los} do not vary systematically with [Fe/H], but undergo a sharp transition at [Fe/H] $\sim$ -0.8. This suggest that there are two discrete populations of clusters, halo and disk, rather than a continuum.

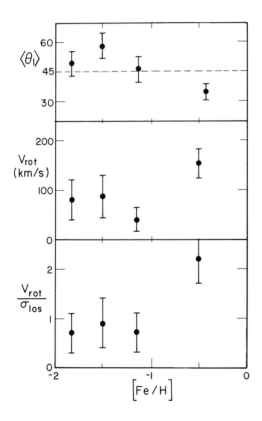

Figure 1.   For groups of globular clusters that have $R \leq R_\odot$, the flattening parameter $\langle\theta_1\rangle$ (in degrees), the rotational velocity, V_{rot}, and the diagnostic of rotational flattening, V_{rot}/σ_{los}, are plotted against the mean value of [Fe/H].

Ages

The advent of CCD detectors has greatly improved the quantity and the quality of data suitable for dating globular clusters. Simultaneously, new theoretical isochrones have been calculated, and they have been translated from the theoretical H-R diagram to the popular photometric systems to make easier the comparisons with observations (VandenBerg and Bell 1985; Green, Demarque, and King 1984). Despite this progress, the dating of globular clusters is stymied by the old problem of how to determine their distance modulii, which is equivalent to the problem of finding the mean luminosity of the RR Lyrae variables and how it varies with metallicity.

Since what one adopts for the dependence of RR Lyrae luminosity on metallicity largely dictates what one finds for the age-metallicity relationship, this problem has a large effect on our understanding of the evolution of the Galaxy. To illustrate this, let us consider, following Buonanno (1986), three popular scenarios for the variation in luminosity. For case 1, it is assumed that the absolute visual magnitude of the RR Lyrae variables is 0.6 for all metallicities (i.e., $M_V^{RR} = 0.6$, $\Delta M_V^{RR}/\Delta[Fe/H] = 0$). Two recent measurements of the luminosities of field RR Lyrae variables (Strugnell et al. 1986; Hawley et al. 1986) by statistical parallax provide some support for this case, which was once widely adopted. These studies, which were based on largely the same data, found $M_V^{RR} = 0.75\pm0.2$ and no evidence for a trend in M_V^{RR} with the spectroscopic metallicity parameter ΔS, although Hawley et al. caution that their solutions for different ΔS groups do not carry much statistical weight. For case 2, it is assumed that $\Delta M_V^{RR}/\Delta[Fe/H] = 0.18$, which is the slope given by theoretical calculations of horizontal branch luminosity (Iben and Renzini 1984) and for comparison with case 1, $M_V^{RR} = 0.6$ at $[Fe/H] = -2$. For case 3, it is assumed again that $M_V^{RR} = 0.6$ at $[Fe/H] = -2$ and that $\Delta M_V^{RR}/\Delta[Fe/H] = 0.35$, which is the variation that Sandage (1982a) found was necessary to explain the observed tight correlation between metallicity and the mean period of the type ab RR Lyrae variables in a cluster (i.e., the Oosterhoff effect). By fitting the main-sequences of 15 globular clusters to each other, Buonanno (1986) has confirmed Sandage's result. It's origin is not readily understood, however, which casts some doubt on its validity. Sandage's explanation that the helium abundances (Y) of the clusters are anticorrelated with $[Fe/H]$ is inconsistent with some observational data (see, for example, Buzzoni et al. 1983) and intuitively seems to be in the wrong sense. If Y does not vary in this way, then the conventional models of horizontal branch stars must be incorrect (see, for example, Renzini 1983). The results of applying the Baade-Wesselink method to field RR Lyrae variables suggest that $\Delta M_V^{RR}/\Delta[Fe/H]$ is positive (see Jameson 1986); hence they support cases 2 and 3 more than 1.

To derive the cluster ages in our illustration, it is convenient to consider the quantity $\Delta M_{TO}^{RR}(bol)$ which is the difference in bolometric magnitude between the main-sequence turnoff point and the zero age horizontal branch (see VandenBerg 1986 for a review of other dating parameters). Because ΔM_{TO}^{RR}(bol) is reddening independent and insensitive to the uncertainty in the mixing length used in stellar models, it is perhaps the best age diagnostic available (see Iben and Renzini 1984). Sandage (1982a) showed that ΔM_{TO}^{RR}(bol) is constant, to within the errors, among clusters spanning the range in metallicity from M92 to M71 (i.e., [Fe/H] = -2.24 to -0.58). More recent data for additional clusters have confirmed this and have yielded a mean value of 3.46 (Buonanno 1986), which for the purposes of our illustration is assumed to hold from [Fe/H] = -2.5 to 0.0. Using Buonanno's (1986) equation 1, which was derived from VandenBerg and Bell's (1985) isochrones with the assumption that Y = 0.23 (Buzzoni et al. 1983), ages for hypothetical clusters of three different metallicities were calculated for each of the above cases, and these results are presented in Table 1. The metallicities were chosen to match approximately the observed spread in metallicity among the halo (-2.5 $\leq$ [Fe/H] $\leq$ -0.8) and disk (-0.8 $\leq$ [Fe/H] $\leq$ 0.0) clusters.

TABLE 1

Ages (in 10^9 yrs) Under Different Assumptions for M_V^{RR}

Case	[Fe/H]		
	-2.5	-0.8	0.0
1	22.6	13.6	10.7
2	20.9	16.3	14.5
3	19.4	19.4	19.4

The data in Table 1 for the halo range in metallicity indicate how radically different is one's picture of the evolution of the Galaxy depending on the choice of RR Lyrae luminosity. According to case 1, the formation of the halo was a very slow process, lasting approximately 40% of the age of the Galaxy. In contrast, according to case 3, the halo formed very rapidly. Predictably, case 2 yields an age-metallicity relation that is intermediate between the extremes of cases 1 and 3.

Since the conclusion that ΔM_{TO}^{RR}(bol) = const. rests on a limited number of data points that have errors of approximately ±0.15 (Buonanno 1986), a small increase or decrease in ΔM_{TO}^{RR}(bol) with increasing [Fe/H], cannot be ruled out. For example, a decrease in ΔM_{TO}^{RR}(bol) that yields for case 3 and age difference of 3 billion years between [Fe/H] = -2.5 and -0.8 is admissible by the existing data. Obviously, it is important to increase the number and the quality of the measurements of ΔM_{TO}^{RR}(bol) to document better its variation with [Fe/H] and, in addition, its variation among clusters of the same [Fe/H]. At present, the scatter at a given [Fe/H] is the same size as that expected from the observational errors alone, but a real difference in age of 2 billion years could be easily hidden by the scatter.

Main-sequence photometry is available for only two of the disk
clusters (M71 and 47 Tuc), which are among the most metal poor of these
clusters. Thus, it is not known whether ΔM_{TO}^{RR}(bol) remains constant
over the metallicity range of the disk system. The true
age-metallicity relation may be quite different from the ones suggested
by the calculations in Table 1. Again it is important to measure more
values of ΔM_{TO}^{RR}(bol), particularly for the very metal rich clusters.

The metallicities of M71 and 47 Tuc are not very different from
the metallicities of the most metal rich of the halo clusters, and
their values of ΔM_{TO}^{RR}(bol) are also not significantly different.
Consequently, under any one of the above cases for M_V^{RR}, M71 and 47 Tuc
are not much younger than these halo clusters. This suggests that the
galactic disk is very old and began to form soon after the halo
completed its collapse or concurrently with the collapse.

Since the galactocentric distances of M71 and 47 Tuc are only ~ 1
kpc less than $R_\odot$, it seems reasonable to assume that the oldest disk
stars in the solar neighborhood are similar to these clusters in age,
which with any reasonable assumption for M_V^{RR} are older than 10×10^9
yrs. This has important implications for the stellar initial mass
function (see Larson this volume).

At present, there is no clear-cut answer to the question of which
of the three cases for M_V^{RR} lies closest to the truth. I believe the
observational evidence favors case 3, which is adopted throughout the
remainder of this paper. It is hoped that data from the HST will solve
this problem once and for all and will yield more precise and many more
values of ΔM_{TO}^{RR}(bol) over the entire range of cluster metallicity. This
will provide a chronology for the formation of the halo and disk and
may also explain the large variations in the morphology of the
horizontal branch that are observed among clusters of the same
metallicity (see below).

Metallicity Gradients

Morgan (1959) and Kinman (1959) showed that the metal-poor
globular clusters are scattered throughout the halo, whereas the
metal-rich ones are found only near the galactic center. Consequently,
if the whole sample of clusters is considered, there is a sizable
gradient in mean metallicity with increasing galactocentric distance.
It may be misleading, however, to infer much about the evolution of the
Galaxy from this observation, for, as we have seen, the cluster system
consists of two separate populations, which undoubtedly formed under
different conditions and possibly at different times. After dividing
the clusters at [Fe/H] = -0.8 into the disk and halo subsystems, it is
less obvious that metallicity gradients exist.

The distances and metallicities of many of the metal-rich clusters
are not accurately know, which precludes definite statements about the

existence of metallicity gradients in the disk system. There is some
marginal evidence for gradients with distance from the galactic center
and plane (see Zinn 1985).

Reasonably precise data exist for 91 halo clusters, and in Figure
2 their values of [Fe/H] are plotted against the logarithm of their
galactocentric distances in kpc. This figure shows that there is a
difference between the metallicity distributions of the clusters within
roughly 7 kpc of the galactic center and those beyond 7 kpc. The
resulting gradient in mean metallicity is not due to the inadvertent
inclusion of a few disk clusters in the R < 7 kpc sample, for the most
metal rich of these clusters have the kinematics of a halo population.
Beyond R $\sim$ 7, there is no sign of a gradient, as Searle and Zinn (1978)
concluded from fewer and less precise measurements. Searle and Zinn's
result was controversial (see Harris and Canterna 1979 and Pilachowski
1984), because the few measurements then available for the very distant
clusters (i.e., R > 60 kpc) suggested that they are more metal poor on
average than the clusters in the $7 \lesssim R \lesssim 40$ kpc zone. The more recent
measurements (see Da Costa, Ortolani, and Mould 1982; Aaronson,
Schommer, and Olszewski 1983; Da Costa 1985; Christian and Heasley
1986), which are included in Figure 2, indicate that the mean
metallicity of these very distant clusters is essentially the same as
the clusters in the 7 - 40 kpc zone. It has been suggested, however,
that the R > 60 clusters may not be members of the same population as
the less distant clusters (see Harris 1976 and Zinn 1985). If they are
excluded from the sample, there is still no evidence for a gradient
beyond R $\sim$ 7.

While there are too few clusters to rule out completely the
existence of a small gradient beyond R $\sim$ 7, this gradient (if one
exists) must be less steep than the one among the R < 7 clusters. The
halo appears to change its characteristics near R $\sim$ 7 kpc, and it may
be more than a coincidence that the R < 7 and the R > 7 groups may
differ kinematically (although the evidence is very weak, see Tables 5
and 7 in Zinn 1985) and that the characteristics of the second
parameter phenomenon also change near R = 7 (see below). It is
disquieting that these changes occur near $R_{\odot}$, but the distance modulii
of the clusters would have to be in error by large amounts for this not
to be the case.

Figure 2 shows that at every radius in the halo, the dispersion in
[Fe/H] is quite large and roughly uniform. This and the absence of a
steep gradient beyond R = 7 are the most significant features of the
diagram. They suggest that the halo collapsed in a chaotic fashion
rather than with a smooth buildup of metals as its radius shrank
(Searle and Zinn 1978).

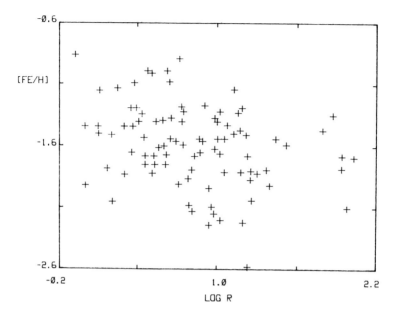

Figure 2. For 91 halo clusters, [Fe/H] is plotted against the logarithm of their galactocentric distances in kpc (data from Zinn 1985 and Christian and Heasley 1986).

The Variation in Horizontal Branch Morphology with R

One of the major results of the investigations of the color–magnitude diagrams of globular clusters is that there is a large variation in the color distribution (i.e., morphology) of the horizontal branch (HB) among clusters of the same metal abundance. For example, the clusters M3, M13, and NGC 7006 have very nearly the same [Fe/H], and yet the HB of M13 consists almost entirely of stars bluer than the RR Lyrae instability strip, that of M3 consists of large numbers of RR Lyrae variables and both redder and bluer stars, and that of NGC 7006 consists of stars redder than the RR Lyrae variables, RR Lyraes, but few bluer stars. It is clear from observations such as these that something besides [Fe/H] influences HB morphology, and this unknown quantity is usually called the second parameter ([Fe/H] is presumably the first).

The morphology of the HB is sensitive to many quantities; consequently there is a long list of potential second parameters (e.g., cluster age; Y; C, N, O abundances; core rotation; see Renzini 1977 for a review). Peterson (1985 and references therein) has provided observational evidence that the surface rotations of HB stars are related to HB morphology, for she has found that the frequency of large rotation rates is largest in clusters that have very blue HB's. It appears unlikely that rotation alone can account for the observed variation in HB morphology with R (see below and Peterson 1985), although it may be the major factor causing several peculiarities of HB morphology (see Freeman and Norris 1981 for a discussion of the need for "non-global" as well as "global" second parameters).

It has been known for some time that the second parameter effect is most prominent among the very distant clusters, and hence its strength must vary with R. Less well known is the fact that there is virtually no evidence for the second parameter among the clusters inside $R_{\odot}$. This is emphasized again here, because it has important implications for the evolution of the Galaxy (see also Searle and Zinn 1978 and Zinn 1980).

In Figure 3, [Fe/H] is plotted against HB type for the clusters in three radial zones of the Galaxy (data from Zinn 1985 and Lee and Zinn 1986). The quantity (B-R)/(B+V+R) is analogous to the parameter B/(B+R) which has been widely used to characterize HB morphology. (B-R)/(B+V+R) has the advantage that the number of RR Lyrae variables is included (B, V, and R are the numbers of blue HB, RR Lyrae, and red HB stars respectively). The top graph of Figure 3 shows that among the R $\leq$ 7 clusters, HB type varies smoothly with [Fe/H] and with very little variation at any given [Fe/H]. Finer subdivisions of this sample by R do not reveal any significant trends of HB type with R. For comparison with the more distant clusters, the region occupied by the R $\leq$ 7 clusters has been "boxed" in and these boxes have been copied in the lower diagrams. The second parameter effect is manifest among the clusters in the 7 < R $\leq$ 40 zone and reaches an extreme in the R > 40 zone. As was the case for the R $\leq$ 7 zone, finer subdivision of the 7 < R $\leq$ 40 zone fails to reveal any significant differences as a function of R.

The tight correlation between HB type and [Fe/H] among the R $\leq$ 7 clusters suggests that in this region of the Galaxy all of the parameters that affect HB morphology are either not varying or are varying in lock step with [Fe/H]. For example, the small scatter at any given value of [Fe/H] is inconsistent with age variations much in excess of 0.5 billion years (Lee and Zinn 1986). This is not true, of course, in the more distant regions, where large variations in one or more of the candidate second parameters is required to explain the observations.

What causes the change at R $\sim$ 7? While this is not known, it seems unlikely that rotation or some abundance anomaly can be the cause. As Searle and Zinn (1978) have argued, these quantities would have to vary very little in the inner regions of the Galaxy and, for some mysterious reason, a lot in the outer regions. The hypothesis that cluster age is the second parameter provides a more attractive explanation. Under this scenario, it is proposed that in the dense inner regions of the proto-Galaxy, metal enrichment proceeded rapidly, and consequently, there was only a brief period of time during which clusters of any particular [Fe/H] could form. The existence of a

metallicity gradient in this region of the halo cluster system suggests
that gaseous dissipation played a role in its collapse, which is
consistent with suggestion that it was a relatively high density
environment. If age is the second parameter, then the clusters in the
7 < R < 40 zone must be younger in the mean than the R < 7 clusters by
about 2 billion years and must have an age spread of roughly 2 billion
years. This can be accounted for in the age scenario by hypothesizing
that the outer halo was built by the destruction of many small systems
consisting of stars, clusters, and gas that dissolved into the halo
over a relatively long period of time. This can also account for the
lack of a metallicity gradient in the outer halo (Searle and Zinn
1978). Finally, it seems reasonable to hypothesize that some of the
small systems remained intact after the collapse of the Galaxy, and
have evolved into the dwarf galaxies that orbit the Galaxy.

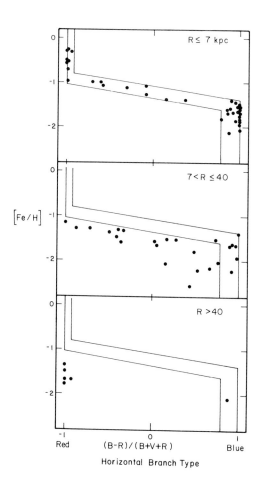

Figure 3. For the globular clusters in three radial zones of the
Galaxy, [Fe/H] is plotted against HB type.

To test this picture, it is necessary to date clusters of the same [Fe/H], but different HB morphology. Unfortunately, the predicted variation in age in the $7 < R \leq 40$ zone is essentially the same as the precision of the very best age determinations, and a definitive test is not possible. The difference in age between the red HB clusters in the $R > 40$ zone and the blue HB clusters in the $R \leq 7$ zone may be 4 billion years and possibly larger. This difference should be measurable in principle, but ground-based observations of the distant clusters simply do not go faint enough to yield very precise ages (see Christian and Heasley 1986). This problem will be much more tractable with the HST.

Comparisons with Field Stars

Evidence that the properties of the globular clusters are not shared by the stars in the galactic field would suggest that they are separate populations, and this would have major implications for the evolution of the Galaxy. Since there is evidence that the globular clusters in giant elliptical galaxies have bluer colors (hence presumably lower metallicities) and more distended spatial distributions than do the field stars in these systems (see Strom et al. 1981; Forte et al. 1981), this is a real possibility. Furthermore, there are theoretical reasons to believe that the formation of globular clusters occurred earlier in the evolution of a galaxy, at a time of low metallicity, than did the formation of individual stars (Fall and Rees 1985). Because the ages of old field stars cannot be determined with much precision, it is not possible to make a meaningful comparison of their ages. Several other properties can be compared, however (see also Norris 1986).

To see if the field stars divide into halo and disk populations at the same metallicity as the globular clusters, Zinn (1985) compared the calculations of V_{rot}, σ_{los}, and V_{rot}/σ_{los} for the clusters with ones for samples of RR Lyrae variables and subdwarfs in the solar neighborhood. The division of the stars at [Fe/H] = -0.8 (i.e., $\Delta S = 2$ and $\delta(U-B) = 0.16$) produced large differences in kinematics which resembled the difference between the halo and disk systems of clusters. Norris (1986) obtained similar results for a larger sample of stars and showed that there is no correlation between metallicity and kinematics among his sample of halo stars, which is in good agreement with the kinematics of the globular clusters (see Figure 1) and is consistent with the absence of a metallicity gradient among the outer halo clusters and the presence of only a small one between the inner and outer halo clusters.

As expected from their kinematics and in good agreement with the clusters, the metal-rich RR Lyrae variables have the spatial distribution of a disk system, for they are concentrated within $\sim$ 3 kpc of the galactic plane (Butler, Kinman, and Kraft 1979; Sandage 1982b). In the past, there was some difficulty with identifying the metal-rich field RR Lyrae variables with the metal-rich clusters, because none of the clusters appeared to contain RR Lyrae variables (see Taam, Kraft, and Suntzeff 1976). The recent discoveries by Hazen-Liller (1985, 1986) of RR Lyrae variables in the metal-rich clusters NGC 6569 and NGC 6388 help alleviate this problem.

The spatial distributions of the halo RR Lyrae variables and the halo clusters also appear to be similar. Saha (1985) has shown that the density of RR Lyrae variables falls off roughly as R^{-3} out to 25 kpc, but falls off as more steeply beyond that point. The decline in density of the globular clusters is similar, including the change in slope near 25 kpc (see Zinn 1985).

The velocity ellipsoid of halo stars has received a lot of attention recently because some studies have suggested that it is different from that of the globular clusters and because different samples of halo stars have given conflicting results (see Hartwick 1983; Pier 1984; Ratnatunga and Freeman 1985; Hartwick and Cowley 1985; Sommer-Larsen and Christensen 1985; Carney and Latham 1986; Norris 1986). It is not clear what weight should be attached to the differences that some of these authors have found between their calculations of the velocity ellipsoid and often quoted conclusion of Frenk and White (1980) that the ellipsoid of the globular clusters is essentially isotropic, which was based on modelling the kinematics. Norris (1986) recently calculated the ellipsoids of large samples of halo stars (selected by non-kinematic criteria to avoid biases) and globular clusters and found that no significant differences exist.

Since metal abundance related criteria are often used to isolate old field stars, it is not a simple matter to make meaningful comparisons of the metallicity distributions of the globular clusters and samples of field stars. For RR Lyrae variables, there is the added complication that their frequency in a stellar population depends on the morphology of its HB, which depends, of course, on metallicity as well as other factors. This selection effect can be used to advantage, however, to test whether the correlation between metallicity and HB type is the same in the field as among the clusters. The color-magnitude diagrams of globular clusters show that this correlation changes with increasing R (see Figure 3). If a similar variation is not found among the halo field stars, then there must be a difference between the clusters and the field in one or more of the parameters that control HB morphology.

This possibility is examined below using the extensive ΔS measurements that have been made of the RR Lyrae variables in the Lick astrograph fields (Butler, Kinman, and Kraft 1979; Butler et al. 1982; Kinman et al. 1985). For comparison with the cluster observations, these measurements have been transformed to Zinn and West's (1984) metallicity scale by the equation: [Fe/H] = $-0.16\ \Delta S - 0.41$, which was derived from the globular clusters in Zinn and West's study that have measured values of ΔS. This equation has the same slope as Butler's (1975) calibration of ΔS in terms of [Fe/H], and over the range

$0 < \Delta S < 12$, it is very similar to Manduca's (1981) calibration for the case that $[Ca/H] = 0.8 \, [Fe/H]$.

As noted above, Figure 3 shows that among the globular clusters in the 7-40 kpc zone there is a wide range in HB type at each metallicity. If HB type and metallicity are as poorly correlated in the field, and if the field has the same metallicity distribution as the globular clusters, then the metallicity distribution of the field RR Lyrae variables should be identical to that of the clusters. The variables in the Lick fields in the directions of the north galactic pole (NGP) and the galactic anticenter lie in this radial zone. Since there are no significant differences between the distributions over ΔS of the variables in these fields (Butler et al. 1982), they have been lumped together in the following analysis. Like the globular clusters in the 7-40 zone, there is no evidence for metallicity gradients with either R or distance from the galactic plane ($|z|$) among these variables (see Kinman et al. 1985).

In Figure 4, generalized histograms (following Searle and Zinn 1978) are plotted for the globular clusters and the RR Lyrae variables in the 7-40 zone. This figure shows that there is very little difference in their metallicity distributions. The small difference at low metallicity, which is not statistically significant, may be a consequence of the fact that very metal-poor variables have somewhat smaller amplitudes than do more metal-rich ones, and hence have a slightly lower probability to be found in variable star surveys.

The third field for which measurements of ΔS exist (field RR-I) lies in the R < 7 zone ($4.1 < R < 6.8$, Kinman et al. 1985). As Figure 3 illustrates, there is a tight correlation between metallicity and HB type among the globular clusters in this zone, and most significantly, that there is a relatively small range in $[Fe/H]$ (roughly -1.0 to -1.6) where are found clusters of intermediate HB type. The frequency of RR Lyrae variables is highest in these clusters. If the field and cluster populations are similar, one expects the metallicity distribution of the field RR Lyrae variables to peak near the middle of this range. The variables observed by Kinman et al. (1985) have a mean ΔS of 5.8, which corresponds to $[Fe/H] = -1.34$ on the metallicity scale used here.

To see if the RR Lyrae variables in the globular clusters are similar to the ones in the RR-I field, I have constructed a generalize histogram of their metallicities from the observed numbers of RR Lyrae variables in the clusters (data from Hogg 1973 and the more recent literature) and the metallicities of the clusters. Since the total number of RR Lyrae variables in a cluster depends on its mass as well as its HB morphology, the number of variables per unit cluster luminosity was calculated by dividing the total number of variables by the luminosity of the cluster in V light. It was assumed that all of the variables in a cluster have the mean metal abundance of the cluster, and to obtain a sufficiently large sample of clusters, it was necessary to pick them from a somewhat larger range in R than is

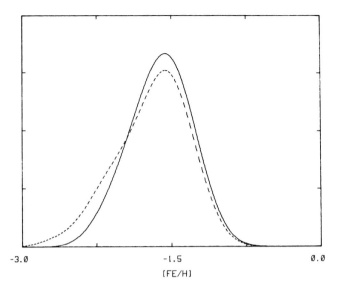

Figure 4. Solid curve: the generalized histogram of the RR Lyrae
variables in the NGP and anticenter fields. Dashed curve: the
histogram of the globular clusters in the 7 – 40 kpc zone. The
ordinate is the relative frequency of the objects. The histograms have
been normalized so that they have the same areas, and a Guassion kernel
with $\sigma([Fe/H]) = 0.2$ was used for both of them.

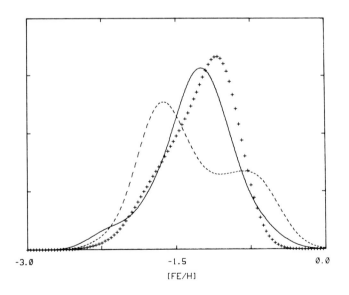

Figure 5. Solid curve: the generalized histogram of the RR Lyrae
variables in field RR-I. Dashed curve: the histogram of the globular
clusters in a similar region of the Galaxy. Crosses: the histogram of
the RR Lyrae variables in these globular clusters (see text).

spanned by the variables in RR-I. The lower limit on the distance from the galactic plane ($|z|$ = 1.1 kpc, Kinman et al. 1985) was kept the same to avoid placing too many disk clusters in the sample.

The results are shown in Figure 5, where are plotted the generalized histograms of the RR Lyrae variables in RR-I, the globular clusters in approximately the same region of the Galaxy, and the RR Lyrae variables (per unit luminosity) in these clusters. One can see that the histogram for the clusters is a poor match to that of the variables in RR-I. The bimodal form of the cluster distribution is a consequence of there being of both halo and disk clusters in the sample (see Freeman and Norris 1981 and Zinn 1985). The histogram for the variables in these clusters show no sign of the two peaks in the cluster distribution. Its single peak lies in fact in the "valley" of the cluster distribution. This shows how strongly HB morphology can affect the metallicity distributions of RR Lyraes. The histogram for the variables in the clusters is a reasonably good match to the one for the variables in RR-I. This suggests that the clusters and the field have similar correlations between [Fe/H] and HB type.

The comparisons in Figures 4 and 5 suggest that the correlation between metallicity and HB branch type undergoes a change in the field, as well as among the clusters, near R = 7 kpc. This change is probably largely responsible for the fact that samples of RR Lyrae variables exhibit larger metallicity gradients with R than do the globular clusters (see Castellani, Maceroni, and Tosi 1983; and Kinman et al. 1985). For example, Kinman et al. found mean values of ΔS of 5.80 ± 0.39 and 7.65 ± 0.22 ([Fe/H] = −1.34 ± 0.06 and −1.63 ± 0.04), respectively, for the variables in RR-I and in the outer halo fields, whereas the halo globular clusters in the zones 3.5-7 kpc and 7-40 kpc have <[Fe/H]> = −1.55 ± 0.07 and −1.69 ± 0.07, respectively (Zinn 1985).

Kraft et al. (1982) have shown that samples of red giants in the globular clusters M13, M15, and M92 have different distributions of [C/Fe] and [N/Fe] than do field red giants of approximately the same [Fe/H]. There is, however, a good correspondence between the field stars and a sample of red giants in the globular cluster M3. On the basis of these observations and similar ones of the red giants in the distant globular clusters Pal 13 and NGC 7006 (Friel et al. 1982), Kraft (1983) has suggested that the halo field stars may be more representative of the red giants in clusters that have red HB's than ones that have blue HB's (the HB's of M3, Pal 13, and NGC 7006 are redder than those of M13, M15, and M92), and he has presented an argument on how this may have occurred, which is based on the hypothesis that core rotation is the second parameter. The comparisons made in Figures 4 and 5 suggest, however, that the second parameter behaves much the same way among the field stars as the clusters, and thus it seems unlikely that the field has a redder HB in the mean than the clusters.

Several authors (see Norris and Pilachowski 1985 and references therein) have argued that the anomalies in C, N, Na, and Al among the red giants in globular clusters may be related to ability of dense clusters to retain the ejecta of highly evolved or dying stars, which presumably have gotten incorporated into later generations of stars in the clusters. Thus, as Kraft (1983) has pointed out, the differences in C and N abundance between some dense clusters and field stars may be a sign that the field stars are the debris of loosely bound clusters. These differences are, therefore, not necessarily evidence that the field stars and the clusters belong to separate populations.

In summary, it appears that the present data on the spatial distributions, kinematics, HB morphologies, and compositions of globular clusters and field stars either suggest that they belong to the same population or leave open this possibility.

OTHER LOCAL GROUP GALAXIES

As noted above, the globular clusters in the other members of the Local Group have been selected on the basis of their appearances without regard to their ages. These clusters should be compared with both the globular and the open clusters in the Milky Way. The color-magnitude diagrams for the star clusters in the Milky Way, LMC, M31, and M33 (data from Harris and Racine 1979; Sagar et al. 1983; Christian and Schommer 1982; van den Bergh 1969, 1981) are compared in Figure 6, which is modeled after figure 3 in Christian and Schommer (1982). Although many more M31 clusters have been measured (see Crampton et al. 1985), the sample plotted in Figure 6 is adequate for our discussion. The effects of interstellar absorption have been removed for the Milky Way clusters. For the other clusters, only the galactic absorption could be dealt with because the absorption that occurs within their parent galaxies has been seldom measured.

Figure 6 illustrates that the luminous (M_V < -6) star clusters in the Milky Way consist of two distinct kinds: the globular clusters, which are redder than (B-V)$_o$ = 0.5, and the much bluer open clusters. At lower luminosity, there is some overlap in color between these two groups. It is important to consider what fraction of these clusters a distant observer would classify as globular, if he were using the same criteria that are used to identify globular clusters in, for example, M31 (e.g., round, non-stellar image). Probably many of the open clusters would be overlooked because they are small in size and because they are disk objects and hence are located in fields of high surface brightness and interstellar absorption. It seems unlikely, however, that all of the open clusters would be missed. In terms of luminosity, color, and size, some of the largest open clusters resemble the blue globular clusters that have been identified in M31 and M33 (see, for example, Schmidt-Kaler 1967). A much larger fraction of the globular clusters would probably be identified, for they are generally large in size, and since they are mostly halo objects, they may lie, depending

on projection effects, in fields of low extinction and surface
brightness. Thus, a distant observer's sample of "globular" clusters
in the Milky Way would probably resemble the sample of M31 clusters, in
that it would contain many more clusters redder than $(B-V)_o$ = 0.5 than
ones bluer than this value.

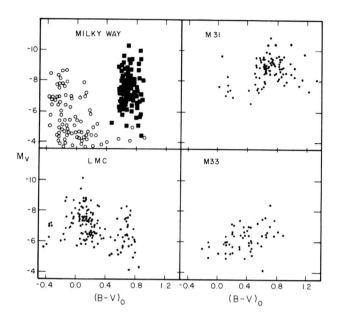

Figure 6. The color-magnitude diagrams for samples of star clusters in
the Milky Way, Large Magellanic Cloud, M31, and M33. Open circles and
filled squares depict, respectively, the open clusters and the globular
clusters in the Milky Way.

The fractions of the total clusters populations that are bluer
than $(B-V)_o$ = 0.5 are much larger in the LMC ($\sim$0.7) and in M33 ($\sim$0.6)
than in M31 ($\lesssim$0.2), Crampton et al. 1985). The fraction of blue
clusters is $\sim$0.8 in the SMC, whose cluster population resembles that of
the LMC (see van den Bergh 1981). Thus, the cluster population of the
late-type spiral M33 (Sc) appears to resemble much more the cluster
populations of the Magellanic irregulars than the earlier type spirals
M31 (Sb) and the Milky Way (Sbc), which was the conclusion reached by
Christian and Schommer (1982, 1983a) and by Cohen, Persson, and Searle
(1984) on the basis of more detailed photometric and spectroscopic
observations.

The fraction of red clusters appears to be related to whether or
not a substantial part of the galaxy's clusters and stars formed during
a halo building phase. The Milky Way and M31 obviously have halos,
whereas it is not clear that even the very oldest clusters in the LMC
are part of a halo population (Freeman, Illingworth, and Oemler 1983).
A few of the reddest clusters in M33 appear to have the kinematics of

halo objects (Christian and Schommer 1983b); hence, in this characteristic M33 may be intermediate between the LMC and the Milky Way. It is attractive to generalize from these few observations that ratio of red to blue clusters is another measure of the bulge to disk ratio and hence varies systematically along the Hubble sequence. Obviously, observations of many more galaxies are required before this can be established. In addition to this general trend, the cluster systems of each of these galaxies have some unique properties, which are briefly discussed below.

M31

The cluster system of M31 is dominated by red clusters that appear to be similar in age to the globular clusters in the Milky Way. The number of these clusters is 3 to 4 times larger in M31 than in the Milky Way, which is consistent with the observations that M31 is more massive than the Milky Way and has a larger bulge to disk ratio. In some other ways, the cluster systems of M31 and the Milky Way are similar, as one might expect since these galaxies are often considered to be near twins. There are a few surprising differences, however, which suggest that these galaxies have evolved in significantly different ways.

In the Milky Way, roughly 25% of the globular clusters belong to the metal-rich disk system. It is controversial whether or not the metal-rich clusters in M31 constitute a disk system. Using the metallicity and radial velocity data published by Huchra et al. (1982) for a sample of 61 clusters, Freeman (1983) found that the velocities of the metal-rich clusters follow the rotation curve of M31's disk, while the metal-poor clusters do not, which he interpreted as evidence for a metal-rich disk and a metal-poor halo population. However, Searle (1986) found no evidence for a metal-rich disk system in his sample of approximately 100 clusters. The absence of a disk system of very old clusters in M31 would suggest that M31's disk is younger than the Milky Way's or, more likely, that the disks are similar in age, but either star clusters did not form in M31's disk at early epochs or they have been destroyed by dynamical processes. It is important to recall that Zinn (1985) found some evidence that the luminosity functions of the halo and disk globular clusters in the Milky Way are different, in the sense that the disk clusters have a broader distribution in luminosity. Furthermore, he found that the disk clusters within 0.5 kpc. of the galactic plane have a lower mean luminosity than the disk clusters lying farther from the plane. It is possible, therefore, that Searle's sample of M31 clusters, which is essentially magnitude limited, simply does not reach faint enough to pick up a sizable population of old disk clusters (Searle 1986).

The measurements of cluster metallicity by van den Bergh (1969) and Huchra, Stauffer and, van Speybroeck (1982) show that there is a shallow gradient in mean metallicity with distance from the center of M31 and that there is a very large dispersion in metallicity at every distance. In these characteristics, the cluster system of M31 closely resembles the halo population of globular clusters in the Milky Way (see Figure 3). However, in M31 the highest metallicity found at a given radius from the center is significantly greatly than that attained by the halo clusters in the Milky Way (van den Bergh 1969, Searle 1983). This appears to be consistent with the larger bulge to disk ratio in M31, which suggests that a larger fraction of its initial mass was converted into stars during its halo building phase (Searle 1979, 1983).

On the basis of these observations it is tempting to conclude that the cluster system of M31 is a scaled-up version of the Milky Way's halo population of globular clusters . That this is not the case has been demonstrated recently by Burstein et al. (1984) who showed that the spectra of some of the brightest of the red clusters in M31 are distinctly different from those of globular clusters in the Milky Way (see also Searle's paper in this volume). Burstein et al. found that the M31 clusters have stronger Balmer lines and CN bands than do Milky Way globulars that have similar strengths of the Mg b - MgH blend at $\lambda \sim 5170$ Å. They were unable to find a completely satisfactory explanation for these differences, although much younger ages (i.e., $\sim 5 \times 10^9$ yrs.) for the M31 clusters appeared to be the most likely of several possibilities.

It is conceivable that these differences between M31 and the Milky Way are related in some way to the anomalies that Rose (1985) and Rose and Tripicco (1986) have discovered among the spectra of the most metal-rich globular clusters in the Milky Way. They found that these clusters divide into two groups according to the surface gravity sensitive index SrII/FeI and CN band strength. They suggested that these differences stem from differences in the giant to dwarf ratio, possibly because of large differences in cluster age.

The observation of the M31 cluster Bo 158 with the IUE satellite by Cacciari et al. (1982) provides yet another hint that the cluster system of M31 may differ substantially from that of the Milky Way. Ground-based observations of Bo 158 indicate that it is a metal-rich cluster, and yet it has a large UV flux, which is characteristic of the metal-poor globular clusters in the Milky Way that have extremely blue HB's. This suggests that the correlation between HB morphology and [Fe/H] is different in the cluster system of M31.

LMC and SMC

The star clusters in the LMC and the SMC span a very large range in age, from approximately the ages of the globular clusters in the Milky Way to less than 10^7 yrs. The distribution over age has been recently constructed by Elson and Fall (1985) for a sample of LMC clusters. The number of clusters was found to decline with increasing age, but more slowly than the number of open clusters in the Milky Way.

In both the LMC and the Milky Way, this decline is probably caused by
dynamical processes, such as evaporation and collisions with molecular
clouds. If the rates of cluster formation have been similar in these
galaxies, then the slower decline in the LMC suggests that these
processes have been less effective in it. Elson and Fall also found no
evidence of bursts of cluster formation, which is interesting because
evidence of bursts of star formation has been seen in some fields of
the LMC (see Frogel and Blanco 1983 and references therein).

Although the ages of the blue clusters in the Magellanic Clouds
are similar to those of the open clusters in the Milky Way, their
masses are substantially larger on average. Their masses and
structures resemble those of the smaller globular clusters in the Milky
Way (Freeman 1980). This raises the question of why have the Clouds
been proficient in forming massive clusters over their entire
histories, while in the disk of the Milky Way, clusters of comparable
mass have not formed since the formation of the old disk globular
clusters.

It is not clear that the very old stars and clusters in the LMC
and the SMC constitute halo populations. The old clusters in the LMC
appear to belong to a disk system that contains objects spanning a
large range in age, from $>10^{10}$ yrs to $\sim 1 \times 10^{9}$ yrs (Freeman et al.
1983). The clusters younger than 1×10^{9} yrs belong to a separate disk
system which also appears to include the HI gas and the HII regions in
the galaxy. Remarkably, these two disks are inclined by $\sim 50°$ to each
other, which suggests that some process has altered the structure of
the LMC in the recent past (Freeman et al. 1983).

The age-metallicity relationships defined by the clusters in the
LMC and the SMC are shown in Figure 7, which is modeled after figure 10
in Stryker, Da Costa, and Mould (1985). Since in some instances the
ages derived from integrated light observations do not agree with ones
based on color-magnitude diagrams (see, for example, Hodge 1983; Searle
1984), only clusters that have measured main-sequence turnoffs have
been plotted in Figure 7. The age-metallicity relationships of the
halo and the disk globular clusters in the Milky Way and the disk stars
in the solar neighborhood are shown schematically. For the globular
cluster relationships, the frequently obtained result that there is no
detectable dispersion in age about a mean value of $\sim 15 \times 10^{9}$ years
(see, for example, Buonanno 1986) has been adopted. Our previous
discussion has cautioned that the true age-metallicity relationships
may be less steep than these results imply. Nevertheless, with any
reasonable estimate of the ages of the globular clusters, it is still
true that the age-metallicity relationship of the halo clusters is
steeper than that of the disk stars. The age-metallicity relationship
of the system of disk globular clusters is particularly uncertain
because only two of its clusters have been dated. While the data for
the LMC and SMC are scanty, they indicate that the age-metallicity
relationships of these galaxies have more nearly the same slope as the
disk stars in the Milky Way than the halo globular clusters. Figure 7

also shows the well-documented fact that the clusters in the Clouds are more metal-poor than the disk stars in the solar neighborhood by considerable factors. Thus, the metal enrichments of the Clouds have occurred slowly, and possibly without the initial spurts produced by the formation of halos.

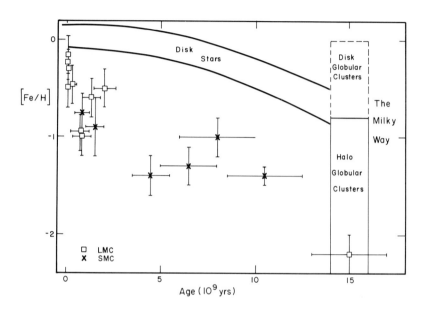

Figure 7. The age-metallicity relationships defined by the star clusters in the LMC and the SMC and by objects in the Milky Way (data from Gascoigne et al. 1981; Rich et al. 1984; Mould et al. 1984; Stryker et al. 1985; Da Costa and Mould 1986; Mould et al. 1986; Da Costa et al. 1985; Cohen 1982; Hodge 1983; Stryker 1983; Twarog 1980 and references therein.

M33

As noted previously, the cluster system of M33 resembles the systems of the LMC and the SMC in that the majority of its clusters are young objects. Unlike the Clouds, however, there is evidence that the oldest clusters in M33 are members of a halo population (Christian and Schommer 1983b). The observations of Christian and Schommer (1983a) and Cohen, Persson, and Searle (1984) indicate that these halo clusters span the range in metallicity from [Fe/H] $\lesssim$ -2 to $\sim$ -1. The age-metallicity relationship of M33's clusters may be intermediate to those of the Milky Way and the Clouds. Like the Milky Way, M33 built a halo, but its halo appears to be somewhat more metal poor. The subsequent evolution of M33's disk population of clusters probably paralleled the slow increase in metals displayed by the disk stars in the Milky Way and the star clusters in the Clouds.

Fornax

The nearest early type galaxy that contains star clusters is the
Fornax dwarf spheroidal galaxy, which contains five clusters that
closely resemble the globular clusters in the Milky Way. Buonanno et
al. (1985) have recently constructed color-magnitude diagrams that
reach the level of the HB in four of these clusters. These diagrams
show that the clusters contain blue as well as red HB stars, and a
number of stars suspected of variability have the right magnitudes of
colors to be RR Lyrae variables. One of the clusters contains a W
Virginis variable (Light et al. 1986), which are found in globular
clusters that have blue HB's. Thus, in terms of HB morphology, these
clusters resemble the metal-poor globular clusters in the Milky Way,
which suggests that they do not have grossly different ages. The
Fornax clusters range in metallicity from [Fe/H] = -2.2 to -1.2, which
suggests that the metal enrichment of Fornax occurred rapidly (see
Buonanno et al. 1985 and references therein).

In several respects, the field population of Fornax does not
resemble its star clusters. It has a higher mean metallicity and a
much redder HB. It also contains populations of carbon stars and main
sequence stars that are much younger than the clusters (see references
in Zinn 1985b). These observations indicate that the field of Fornax
has experienced a long history of star formation. The star clusters of
Fornax do not give, therefore, a complete picture of its evolution.
While this may be nothing more than a consequence of the small size of
Fornax, which limited its cluster population to just a few objects, it
does serve as a warning that one should be cautious when drawing
conclusions about the evolution of a galaxy from observations of only
its star clusters.

This research was supported by NSF grant AST-8304034.

REFERENCES

Aaronson, M., Schommer, R. A. & Olszewski, E. W. 1984, Astrophys. J.
    276, 221.
Baade, W. 1958, in Stellar Populations, ed. D. J. K. O'Connell
    (Amsterdam: North Holland), p. 303.
Binney, J. 1978, Mon. Not. Roy. Astr. Soc., 183, 501.
Buonanno, R., Corsi, C. E., Fusi Pecci, F., Hardy, E., & Zinn, R. 1985,
    Astr. Astrophys. 152, 65.
Buonanno, R. 1986, Mem. Soc. Astr. Italiana, in press.
Burstein, D., Faber, S. M., Gaskell, C. J., & Krumm, N. 1984,
    Astrophys. J., 287, 586.
Butler, D. 1975, Astrophys. J., 200, 68.
Butler, D., Kinman, T. D. & Kraft, R. P. 1979, Astron. J., 84, 993.

Butler, D., Kemper, E., Kraft, R. P. & Suntzeff, N. B. 1982, Astron.
    J., 87, 353.
Buzzoni, A., Fusi Pecci, F., Buonanno, R., & Corsi, C. E. 1983, Astr.
    Astrophys., 128, 94.
Cacciari, C., Cassatella, A., Bianchi, L., Fusi Pecci, F. & Kron, R. G.
    1982, Astrophys. J., 261, 77.
Carney, B. W. 1984, Pub. Astr. Soc. Pac., 96, 841.
Carney, B. W. & Latham, D. W. 1986, Astron. J. 92, 60.
Castellani, V., Maceroni, C. & Tosi, M. 1983, Astr. Astrophys., 128,
    64.
Christian, C. A. & Schommer, R. A. 1982, Astrophys. J. Suppl., 49, 405.
Christian, C. A. & Schommer, R. A. 1983a, Astrophys. J., 275, 92.
Christian, C. A. & Schommer, R. A. 1983b, in IAU Symp. 100, Internal
    Kinematics and Dynamics of Galaxies, ed. E. Athanassoula, (Reidel,
    Dordrecht), p. 365.
Christian, C. A. & Heasley, J. N. 1986, Astrophys. J., 303, 216.
Cohen, J. G. 1982, Astrophys. J., 258, 143.
Cohen, J. G., Persson, S. E., & Searle, L. 1984, Astrophys. J., 281,
    141.
Crampton, D., Cowley, A. P., Schade, A. & Chayer, P. 1985, Astrophys.
    J., 288, 494.
Da Costa, G. S., Ortolani, S., & Mould, J. 1982, Astrophys. J. 257,
    633.
Da Costa, G. S. 1985, Astrophys. J. 291, 230.
Da Costa, G. S., Mould, J. R., & Crawford, M. D. 1985, Astrophys. J.,
    297, 582.
Da Costa, G. S. & Mould, J. R. 1986, Astrophys. J., 305, 214.
Elson, R. A. W. & Fall, S. M. 1985, Astrophys. J., 299, 211.
Fall, M. & Rees, M. 1985, Astrophys. J. 298, 18.
Forte, J. C., Strom, S. E., & Strom, K. M. 1981, Astrophys. J., 245,
    L9.
Freeman, K. C. 1980, in IAU Symp. 85, Star Clusters, ed. J. E. Hesser,
    (Reidel, Dordrecht), p. 317.
Freeman, K. C. & Norris, J. 1981, Ann. Rev. Astr. Ap., 19, 319.
Freeman, K. C. 1983, in IAU Symp. 100, Internal Kinematics and Dynamics
    of Galaxies, ed. E. Athanassoula, (Reidel, Dordrecht), p. 359.
Freeman, K. C. Illingworth, G. & Oemler, A. Jr. 1983, Astrophys. J.,
    272, 488.
Frenk, C. S. & White, S. D. M. 1980, Mon. Not. Roy. Astr. Soc., 193,
    295.
Frenk, C. S. & White, S. D. M. 1982, Mon. Not. Roy. Astr. Soc., 198,
    173.
Frogel, J. A. & Blanco, V. M. 1983, Astrophys. J., 274, L57.
Friel, E., Kraft, R. P., & Suntzeff, N. B. 1982, Pub. Astron. Soc. Pac.
    94, 873.
Gascoigne, S. C. B., Bessell, M. S., & Norris, J. 1981, in IAU Coll. 68
    Astrophysical Parameters for Globular Clusters, eds. A. G. D.
    Philip & D. S. Hayes (Davis, Schenectady). p. 365.
Green, E. M., Demarque, P., & King, C. R. 1984, Bull. Am. Astron. Soc.,
    16, 997.
Harris, W. E. 1976, Astron. J., 81, 1095.

Harris, W. E. & Canterna, R. 1979, Astrophys. J., 231, L19.
Harris, W. E. & Racine, R. 1979, Ann. Rev. Astr. Ap., 17, 241.
Hartwick, F. D. A. 1983, Mem. Soc. Astr. Italiana, 54, 51.
Hartwick, F. D. A. & Cowley, A. P. 1985, Astron. J. 90, 2244.
Hawley, S. L., Jeffreys, W. H., Barnes, T. G., & Lai, W. 1986,
    Astrophys. J. 302, 626.
Hazen-Liller, M. L. 1985, Astron. J., 90, 1807.
Hazen-Liller, M. L. 1986, preprint.
Hesser, J. E. 1983, Mem. Soc. Astr. Italiana, 54, 7.
Hodge, P. W. 1983, Astrophys. J., 264, 470.
Hogg, H. S. 1973, Pub. David Dunlap Obs., vol. 3, no. 6.
Huchra, J., Stauffer, J., & Van Speybroeck, L. 1982, Astrophys. J.,
    259, L57.
Iben, I. Jr. & Renzini, A. 1984, Phys. Reports, 105, 329.
Jameson, R. F. 1986, Vistas Astr., 29, 17.
Kinman, T. D. 1959, Mon. Not. Roy. Astr. Soc. 119, 538.
Kinman, T. D., Kraft, R. P., Friel, E., & Suntzeff, N. B. 1985,
    Astron.J., 90, 95.
Kraft, R. P., Suntzeff, N. B., Langer, G. E., Carbon, D. F., Trefzger,
    Ch. F., Friel, E., & Stone, R. P. S. 1982, Pub. Astron. Soc. Pac.,
    94, 55.
Kraft, R. P. 1983, in Highlights of Astronomy Vol. 6, ed. R M. West
    (Reidel, Dordrecht), p. 129.
Lee, Y.-W. & Zinn, R. 1986, in preparation.
Light, R. M., Armandroff, T. E. & Zinn, R. 1986, Astron. J., 92, 43.
Manduca, A. 1981, Astrophys. J., 245, 258.
Morgan, W. W. 1959, Astron. J., 64, 432.
Mould, J. R., Da Costa, G. S. & Crawford, M. D. 1984, Astrophys. J.
    280, 595.
Mould, J. R., Da Costa, G. S., & Crawford, M. D. 1986, Astrophys. J.,
    304, 265.
Norris, J. & Pilachowski, C. A. 1985, Astrophys. J., 299, 295.
Norris, J. 1986, preprint.
Peterson, R. C. 1985, Astrophys. J., 294, L35.
Pier, J. R. 1984, Astrophys. J., 281, 260.
Pilachowski, C. A. 1984, Astrophys. J., 281, 614.
Ratnatunga, K. U. & Freeman, K. C. 1985, Astrophys. J. 291, 260.
Renzini, A. 1977, in Advanced Stages in Stellar Evolution, 7th Advanced
    Course Saas-Fee, Switzerland (Geneva Obs., Sauverny). p. 151.
Renzini, A. 1983, Mem. Soc. Astr. Italiana, 54, 335.
Rich, R. M., Da Costa, G. S. & Mould, J. R. 1984, Astrophys. J., 286,
    517.
Rose, J. A. 1985, Astron. J., 90, 1927.
Rose, J. A. & Tripicco, M. J. 1986, preprint.
Sagar, R., Joshi, U. C., & Sinvhal, S. D. 1983, Bull. Astr. Soc. India,
    11, 44.
Saha, A. 1985, Astrophys, J., 289, 310.
Sandage, A. 1982a, Astrophys. J. 252, 553.
Sandage, A. 1982b, Astrophys. J. 252, 574.
Schmidt-Kaler, Th. 1967, Astron. J. 72, 526.
Searle, L. & Zinn, R. 1978, Astrophys. J., 225, 357.

Searle, L. 1979, in Les Elements et Leurs Isotopes Dans L'Univers,
    Liege coll 22, p. 437.
Searle, L. 1983, Carnegie Inst. of Wash. Yearbook 82, p. 622.
Searle, L. 1984, in IAU Symp. 108, Structure and Evolution of the
    Magellanic Clouds, eds. S. van den Bergh & K. S. de Boer (Reidel,
    Dordrecht), p. 13.
Searle, L. 1986, private communication.
Sommer-Larsen, J. & Christensen, P. R. 1985, Mon. Not. Roy. Astr. Soc.,
    212, 851.
Strom, S. E., Forte, J. C., Harris, W. E., Strom, K. C., Wells, D. C.,
    & Smith, M. G. 1981, Astrophys. J., 245, 416.
Strugnell, P., Reid, N. & Murray, C. A. 1986, Mon. Not. Roy. Astr.
    Soc., 220, 413.
Stryker, L. L. 1983, Astrophys. J. 266, 82.
Stryker, L. L., Da Costa, G. S. & Mould, J. R. 1985, Astrophys. J.,
    298, 544.
Taam, R. E., Kraft, R. P., & Suntzeff, N. 1976, Astrophys. J., 207,
    201.
Twarog, B. A. 1980, Astrophys. J., 242, 242.
VandenBerg, D. A. & Bell, R. A. 1985, Astrophys. J. Suppl. 58, 561.
VandenBerg, D. A. 1986, Mem. Soc. Astr. Italiana, in press.
van den Bergh, S. 1969, Astrophys. J. Suppl., 19, 145.
van den Bergh, S. 1981, Astron. Astrophys. Suppl., 46, 79.
Zinn, R. 1980, Astrophys. J., 241, 602.
Zinn, R. & West, M. J. 1984, Astrophys. J. Suppl. 55, 45.
Zinn, R. 1985a, Astrophys. J., 293, 424.
Zinn, R. 1985b, Mem. Soc. Astr. Italiana, 56, 223.

DISCUSSION

KING: Where do the field RR Lyraes fit in?

ZINN: The field RR Lyrae variables appear to belong to the same
populations as the globular clusters. The variables that have $\Delta S < 2$
belong to the disk system, whereas those with $\Delta S > 2$ belong to the halo
population.

MOULD: You mentioned the need to find unbiassed tracers of the
metallicity distribution in the halo. Two classes of stars come to
mind. 1) K giants, which all low mass stars become, were used by
Whitford and Rich for this purpose. The problem with K giants is that
it is hard to determine their distances, 2) main sequence turn-off
stars, which Gilmore has been using to map the halo. Main sequence
stars are faint, but accurate observations of their broad band colors
at large distances would tell us much of what we want to know in
comparing clusters and field metallicity gradients.

LAIRD: In regard to the metallicity distribution function, we are
currently determining metallicities for the stars in the Carney-Latham
survey. We compare the spectra used for the radial velocities to a
grid of synthetic spectra. The metallicity distribution function for
the resulting large sample of high-velocity dwarfs will be finished
soon.

SANDAGE: What is the evidence that M33 has a halo? You classify a
fraction of the M33 clusters on halo objects and some as disk. What is
the basis of this discussion in M33.

ZINN: These conclusions are based on Christian and Schommer's work who
have shown that the reddest clusters in M33 have radial velocities
quite unlike the HI disk, whereas the velocities of the bluer clusters
closely resemble the disk.

SCHOMMER: A comment on some new M33 cluster results which will be
reported on more fully at the Shapley Symposium. Christian and
Schommer have integrated BVI photometry on 80 additional clusters,
bringing the total with BV photometry to 120 clusters. Allan Sandage
has a catalogue, which Carol Christian has been working on, which
brings the total number of clusters in M33 to about 500 objects. Many
of these are blue, young disk systems, and the cluster system of M33
resembles that of the Magellanic Clouds, in particular the SMC, much
more closely than that of the Milky Way or M31. There are about 15-20
true globulars, based upon integrated colors and spectra. About 15
clusters have radial velocities with measured errors ≤ 40 km s^{-1}. Six
of these are blue (B-V < 0.5) and rotate with the disk HI at their
projected positions (to within the velocity accuracies). Nine red
clusters have velocities typically 50-100 km s^{-1} different than
projected disk velocities, and their 1 dimensional velocity dispersion
is ~ 86 km s^{-1}.

SMITH: How firm is the metallicity scale for the metal-rich globular clusters?

ZINN: The metallicity scale is now on firmer ground than it was just a few years ago.   I believe most workers now believe that the clusters M71 and 47 Tuc are indeed more metal rich than clusters such as M5, as was always indicated by their color-magnitude diagrams, integrated light, and photometry and low resolution spectroscopy of their member stars.   The analyses of high dispersion echelle spectrograms that gave contrary results are now widely believed to have been in error because incorrect placement of the continua.   For metallicities above those of M71 and 47 Tuc, one must extrapolate, and this involves, of course, additional uncertainty.

BURSTEIN: a)   I am a member of one of two groups that have been taking IUE spectra of M31 clobular clsuters.   To the best of my knowledge, none of the 6 or so other M31 globular clusters have shown strong (or any!) Far-UV Emission like BO 158.   I now wonder whether the observation of BO 158 is correct.
b)   You referred to a "problem" with CN in the M31 globular cluster system.   It is worthwhile remembering that this "problem" is: <u>every</u> giant star above the clump and at the clump would have to have a CN strength equivalent to $C(41-42) = 0.5-0.6$, i.e. as strong as the strongest-lined giant observed in our own galaxy.

CACCIARI: I would like to comment on David Burstein's remark that the UV spectrum of BO 158 (i.e. SWP 10280 taken by IUE) might be a spurious effect.   The spectrum is of course extremely faint, but it is clearly visible.   The estimated flux level, obtained with two independent methods of data reduction, one of which was used for very faint QSO spectra, is arount $0.5-1.5 \times 10^{-15}$ ergs cm^{-2} sec^{-1} A^{-1} in the wavelength range $\lambda\lambda 1300-1900$.   The only spurious effect I can think of is a residual from previous overexposures, due to phosphor decay in the UV to optical converter.   The SWP images taken with the previous 16 hours were not overexposed, however according to Snijders, ESA-IUE Newsletter 16, p. 10, also an accumulation of optimum exposures (say a 200 DN exposure every hour for 7-8 hours) can leave a very faint 2.5 to 5 DN peak signal in the full-shift exposure following immediately afterwards.   We analyzed the previous images and found that only 3 had an exposure level around 200 DN in the previous 16 hours, so we felt confident that the faint spectrum visible in SWP 10280 was real.

BURSTEIN: For spurious effect I meant a contribution from the background, i.e. spiral arm of M31.   To support this, the SWP spectrum looks a bit fuzzy, like from an extended source.

CACCIARI: It is true that the SW spectrum looks a bit fuzzy, but not enough to fill the aperture, as it would be for an extended source. The fuzziness could be due to non-optimum focussing or guiding - the LW spectrum in fact is quite sharp and point-source-like.   I find it difficult to imagine a diffuse source which radiates in the range

$\lambda\lambda1000-2000$ Å and does not in the range $\lambda\lambda2000-3000$ Å. On the other hand one never knows, you might be right – the only way to find out is to observe BO 158 again, possibly with another instrument, say Space Telescope?!

THE INITIAL MASS FUNCTION

Richard B. Larson

Yale Astronomy Department, Box 6666,
New Haven, Connecticut 06511, U.S.A.

INTRODUCTION

A knowledge of the mass spectrum with which stars form, and of its possible variations with time and location, is an essential requirement for understanding the stellar content and evolution of galaxies. Nevertheless uncertainties still exist even in the qualitative form of the stellar initial mass function (IMF), and they affect nearly all aspects of galactic evolution. This review will summarize some of the present evidence on the IMF and its variability, some theoretical considerations that may help to explain the observations, and some implications of possible variations in the IMF for understanding the properties of galaxies. In particular, the possibility that massive stars and their remnants play a more important role than in conventional models will be discussed.

OBSERVATIONAL EVIDENCE

An extensive review of the observational evidence on the IMF has recently been given by Scalo (1986), who has compiled a large amount of data and discussed the many uncertainties. Some of the important results will be summarized here; Scalo's review should be consulted for further information and references.

Direct Evidence: Luminosity Functions

The main sequence for nearby stars (Philip & Upgren 1983) is populated with a density of stars that increases monotonically toward fainter magnitudes with two exceptions: there is a marked drop in stellar numbers for magnitudes fainter than about $MV = +13$, and there is a smaller deficiency of stars near $MV = +7$. Both of these features have shown up repeatedly in a variety of studies and are generally considered to be real. For stars fainter than $MV > +6$ or less massive than about $0.85~M\odot$, stellar lifetimes exceed the age of the Galaxy and the initial mass function integrated over galactic history can be obtained directly from the observed luminosity function, given an assumed mass-luminosity relation. If a mass-luminosity relation drawn smoothly through the available data is used, the resulting IMF for low-mass stars shows the same features as the luminosity function, namely a steep decline with

decreasing mass below about 0.25 M⊙, and a small dip at 0.7 M⊙ followed
by an upturn toward 0.85 M⊙. Significant changes in the assumed
mass-luminosity relation would be required for these features in the
derived mass function to be removed or dismissed as unreal, but this
possibility cannot presently be excluded.

A question of major interest is whether enough stars of very low
mass (less than 0.1 M⊙) can be present to account for the unseen mass in
the solar neighborhood. This would require the IMF to increase fairly
steeply with decreasing mass at low masses; however, even a generous
estimate of the uncertainty in the mass-luminosity relation for the
faintest stars does not yield a mass function that increases with
decreasing mass. A straightforward extrapolation of the IMF to lower
masses actually suggests very little mass in unobserved faint stars
(Tinsley 1981b). Of course, it cannot be excluded that the IMF rises
steeply for unobservably faint stars, but at present no evidence or
theoretical argument suggests this.

For masses above 0.85 M⊙, stellar lifetimes are shorter than the
age of the Galaxy and the IMF integrated over galactic history cannot be
obtained from the data without making an assumption about the past star
formation rate (SFR) for stars of each mass as a function of time. The
simplest possibility, first considered by Salpeter (1955), is that the
SFR for stars of all masses has been constant throughout galactic
history. With this assumption, Salpeter found that the IMF for stars
between 0.4 and 10 M⊙ could be represented by a power law $dN/d \log m \propto m^{-x}$ with x = 1.35. Within the (surprisingly large) uncertainties in the
derivation of the IMF from star counts, the IMF derived by Scalo (1986)
for a constant SFR is still adequately represented by Salpeter's power
law between these mass limits, except possibly for the small dip at
0.7 M⊙. However, the IMF of nearby stars clearly falls below the
Salpeter power law at masses below 0.4 M⊙; the IMF derived by Scalo
(1986) reaches a maximum at about 0.25 M⊙ and then drops steeply toward
lower masses. Also, most determinations of the IMF for massive stars
yield a steeper slope than the Salpeter law, although the IMF still has
approximately a power-law form.

Studies of the luminosity functions of the brighter stars in open
clusters have shown that, although some possible variations exist, the
results are generally consistent with the field star IMF for masses
above a few solar masses. A universal form for the upper IMF is also
suggested by recent results for the luminosity function of the brightest
stars in a number of nearby galaxies (Freedman 1985); within the
uncertainties, the luminosity function has the same form in all of the
galaxies studied. It appears that most, if not all, of the available
star count data for stars more massive than a few solar masses are
consistent with a nearly universal power-law form for the upper IMF with
a slope x = 1.8 ± 0.5 (Scalo 1986).

The apparent dip at 0.7 M⊙ in the IMF of nearby field stars
suggests that the IMF may not vary monotonically with mass for masses
near 1 M⊙. The form of the local IMF between 0.85 M⊙ and about 2 M⊙

cannot be determined directly from the data because it depends strongly on the unknown past history of star formation. If the SFR has been constant with time, the deduced IMF rises from the dip at 0.7 M0 to a small bump near 1 M0; if the SFR has decreased with time, as is more plausible astrophysically because of the steady depletion of the available gas supply, the size of the bump increases and the conclusion becomes unavoidable that the IMF is bimodal and has a second peak at a mass somewhat above 1 M0 (Scalo 1986, Larson 1986a). Although it is not definitely established by these considerations that the IMF in the solar neighborhood is bimodal, the "continuity constraint" that models of galactic evolution should yield a monotonic IMF (Miller & Scalo 1979) no longer seems very compelling; models with a declining SFR and a nonmonotonic IMF are not excluded, and may even be more compatible with the star count data (Armandroff 1983).

A further suggestion that star formation may, at least in some locations, produce an IMF that peaks at a mass above 1 M0 is provided by the fact that the luminosity functions of some open clusters appear to be deficient in stars less massive than 1 M0 (Scalo 1986). van den Bergh (1972) suggested on this basis that star formation is bimodal, field stars being formed by a low-mass mode of star formation and open cluster stars by a high-mass mode having an IMF that peaks at a mass somewhat above 1 M0. Present data do not suggest that a deficiency of low-mass stars is a general property of all open clusters, but it nevertheless appears that there are large variations in the IMF of low-mass stars in open clusters. Large variations have also been found in the luminosity functions of Magellanic Cloud clusters (Freeman 1977) and galactic globular clusters (McClure et al., this conference), so it seems established that, at least in localized regions, large deviations from a Salpeter IMF can occur for intermediate- and low-mass stars. The variability of the lower IMF and apparent universality of the upper IMF may be explainable if star formation always produces a peaked IMF with a power-law tail toward high masses but the mass at which the IMF peaks is different in different regions.

Indirect Evidence: Galactic Evolution Models

Indirect evidence concerning the form of the local IMF and its possible variability with time may be provided by galactic evolution models constructed to solve two classical problems: the nature of the unseen mass in the solar vicinity (Oort 1965, Bahcall 1986) and the relative paucity of metal-poor stars (Schmidt 1963). Closely related to the latter problem is that of explaining the stellar age-metallicity relation (Twarog 1980, Carlberg et al. 1985), which shows only a modest variation of metallicity with time over most of galactic history.

Between one-half and three-quarters of the dynamically determined mass in a column through the galactic disk at the solar location has not yet been identified with visible stars or gas (Bahcall 1986). Conventionally, it has been assumed that this local "missing mass" is in very dim low-mass stars, but we have seen that there is little evidence

supporting this assumption. If the unseen mass is not in low-mass
stars, the only reasonable alternative is that it is in stellar
remnants, since the unseen matter is confined to a disk (Bahcall 1986)
and must therefore once have consisted of gas that condensed into
star-like objects that are now unseen.

The column density of the unseen matter in the solar neighborhood
is between about 30 an 90 $M\Theta/pc^2$, depending on its scale height; numbers
in the lower part of this range are appropriate if the unseen matter has
a scale height comparable to that of the old disk stars. A model
assuming a standard monotonic IMF and a constant SFR over galactic
history predicts a remnant column density of about 10 $M\Theta/pc^2$ (Tinsley
1981b), but the predicted mass in remnants can be larger than this if
the SFR has decreased with time and the IMF is not monotonic (Quirk &
Tinsley 1973). In such models the past average star formation rate can
easily be three or more times the present rate (Armandroff 1983, Scalo
1986, Larson 1986a); the predicted remnant column density then becomes
30 $M\Theta/pc^2$ or more, enough to account for the unseen mass. If the IMF
has remained unchanged in form while the SFR has decreased, the derived
IMF in these models has a second peak at a mass of 1.2 to 1.5 $M\Theta$.

If the IMF is the sum of high-mass and low-mass components, it is
possible that the two components have not varied in the same way with
time. Only the high-mass mode of star formation is required to have a
decreasing SFR in order to account for the unseen mass as remnants; the
low-mass mode may form stars at a nearly constant rate, as the
observations suggest for stars of near solar mass (Twarog 1980). A
model with a bimodal IMF of this type that is consistent with all of the
data and accounts for the unseen mass as remnants has been suggested by
Larson (1986a). In this model the IMF has a second peak near 2 $M\Theta$
which, integrated over galactic history, accounts for nearly as many
stars and about five times as much mass turned into stars as the
low-mass peak at 0.25 $M\Theta$.

Like the earlier model of Schmidt (1963) in which the proportion of
massive stars in the IMF is larger at earlier times, this model has the
additional advantage that the stellar age-metallicity relation is
satisfactorily accounted for without further assumptions such as gas
infall. Schmidt's model was devised to account for the stellar
metallicity distribution, but if Schmidt's assumptions about the
variation of the IMF with time are combined with current data (Scalo
1986) for the stellar luminosity function and stellar lifetimes, the
column density of remnants predicted by Schmidt's model is about 50
$M\Theta/pc^2$, even more than the 37 $M\Theta/pc^2$ in Larson's (1986a) model and
easily enough to account for the unseen mass in the solar neighborhood.
Thus, models in which high-mass star formation was more important in the
past provide an appealing possible description of galactic evolution,
since very similar assumptions can account for both the unseen mass and
the stellar age-metallicity relation.

The usual alternative way of accounting for the stellar
age-metallicity relation is to invoke gas infall throughout galactic

history (e.g. Twarog 1980). However, there is little evidence for the
postulated fairly high rates of gas infall; a recent estimate for our
Galaxy (Mirabel & Morras 1984) is that the total infall rate is only
about 0.2 M☉/yr, nearly an order of magnitude smaller than required in
infall models of chemical evolution. A more elaborate model with infall
that may still be compatible with all of the data postulates a slow
radial inflow of gas in the galactic disk (Lacy & Fall 1985); the
required gas flow rates in this model are plausible and do not conflict
with any observed limits.

Indirect Evidence: Integrated Radiation Measures

    Constraints on the IMF in regions of star formation or in galaxies
that are too heavily obscured or too distant for star count studies to
be possible can be obtained from measurements of their integrated
emission at various wavelengths. For example, radio or optical
observations of ionized gas in star forming regions can be used to infer
the ionization rate and hence the number of ionizing stars present, i.e.
the number of O stars with masses greater than about 15 M☉. This may be
compared with the total luminosity at all wavelengths, which comes
mainly from B stars, or with dynamical determinations of the total mass
present, which is mostly in low-mass stars or remnants, to place
constraints on the initial mass function. An additional argument
constraining the IMF may be obtained by comparing the inferred star
formation rates in galaxies with plausible limits on the rate of
depletion of the observed gas supply.

    For the inner Milky Way, Güsten & Mezger (1983) have used the
thermal radio emission from HII regions to derive the surface density
and the formation rate of massive stars as a function of distance from
the galactic center. If a conventional monotonic IMF is adopted to
derive the total star formation rate per unit area, the resulting values
are too high to be easily reconciled with simple models of galactic
evolution; even if the SFR has been constant with time, the predicted
amount of mass in stars and remnants is more than is allowed by the
galactic rotation curve. Güsten & Mezger (1983) therefore suggested
that the IMF in the inner Milky Way favors massive stars, and they
showed that all of the data on star formation and chemical abundance
gradients in our Galaxy can be accounted for if the IMF is bimodal and
has a second peak at about 2 - 3 M☉ whose relative amplitude decreases
with distance from the galactic center. A similar conclusion had
earlier been reached by Jensen, Talbot, & Dufour (1981) from optical
observations of the nearby bright spiral M83; again it was found that
the formation rate of massive stars is too high to be easily reconciled
with standard assumptions, and it was suggested that the IMF in M83
peaks at a mass of at least a few solar masses.

    For the giant HII regions in the Large Magellanic Cloud, a
comparison of the inferred ionization rate with the total luminosity
emitted at all wavelengths indicates an excess of O stars (M > 15 M☉)
compared to B stars by about a factor of two compared with bright Milky

Way HII regions and a factor of four compared with a conventional IMF
(Jones et al. 1986), suggesting again that the formation of massive
stars is favored in regions of active massive star formation.

Even larger departures from a conventional IMF may occur in
starburst galaxies where exceptionally rapid star formation is presently
taking place. A nearby prototype is M82, whose high infrared luminosity
indicates a high rate of formation of massive stars. Rieke et al.
(1980) have used observations at a variety of wavelengths to constrain
the IMF in M82, and they have argued that the data can be explained only
if the IMF is strongly deficient in stars less massive than about 3 M$\odot$;
otherwise not enough luminosity is produced at all of the observed
wavelengths, given the allowed limits on the mass. Similarly,
Viallefond & Thuan (1984) have noted that the IMF in I Zw 36 may lack
stars below about 4 M$\odot$, and Augarde & Lequeux (1985) have found that the
spectrum of Mk 171 permits few stars less massive than several solar
masses and have suggested that the IMF in this galaxy may have a lower
mass limit as high as 10 or 20 M$\odot$. A further indication that star
formation in starburst galaxies probably does not produce many low-mass
stars is that if this were the case, the available gas supply would be
exhausted in an implausibly short time; for example, in M82 the gas in
the starburst region would be exhausted in less than 10^7 yr (Kronberg,
Biermann, & Schwab 1985), a conclusion that would conflict with evidence
that star formation in M82 has been going on for longer than 10^7 yr.
Thus it seems likely that the IMF in starburst galaxies does not contain
the standard proportion of low-mass stars.

A similar, although less severe, problem of rapid gas depletion is
encountered widely in star-forming galaxies (Larson, Tinsley, & Caldwell
1980; Kennicutt 1983). A possible solution involves gas replenishment
by infall, but there is little direct evidence for this. Another
solution, discussed by Sandage (1986), is to postulate that star
formation is bimodal and that the IMF generally contains a smaller
proportion of low-mass stars in relation to the observable massive stars
than does the Salpeter function; the existence of many late-type
galaxies that still contain substantial amounts of gas and young stars
then becomes easier to understand.

Summary of the Evidence

The foregoing brief overview of some of the evidence suggests the
following general features of the IMF that call for theoretical
explanation: (1) Most of the evidence based on star counts is
consistent with a universal, approximately power-law form for the IMF of
massive stars with x $\sim$ 1.8 ± 0.5, somewhat steeper than the Salpeter
law. (2) Many observations suggest a peak or turnover in the IMF at a
low or moderate mass that varies with location. In the solar
neighborhood the main peak in the IMF is at a mass of about 0.25 M$\odot$; in
regions of more active star formation the IMF may have a peak around
2 M$\odot$; and in starburst systems the IMF may peak at an even larger mass,
possibly as high as 10 M$\odot$ or more. All of the data seem consistent with

a peak mass that increases with the star formation rate (Scalo 1986).
(3) It is possible that the IMF in galaxies is generally bimodal or
double-peaked; this could reflect either two mechanisms of star
formation or a bimodal distribution of the parameters controlling the
IMF. Bimodality could, for example, arise from differences between star
formation in spiral arms and star formation in interarm regions (Mezger
& Smith 1977, Larson 1977).

THEORETICAL CONSIDERATIONS

     Our present understanding of star formation is still too
rudimentary to predict a detailed initial mass function, but some recent
theoretical results on fragmentation may bear at least a
semi-quantitative comparison with the observations, and may help to
explain the qualitative trends noted above. We first consider the
mechanisms that might in principle produce a spectrum of masses of
protostellar objects.

Possible Mechanisms for Producing an IMF

     A mass spectrum can be created either by a fragmentation or
"top-down" process that starts with a large cloud and breaks it into
small clumps, or by an accumulation or "bottom-up" process that starts
with small objects and builds up larger ones by coagulation or
accretion. Both types of process may play a role in establishing the
stellar initial mass function: fragmentation processes, which might be
expected to depend on cloud properties, may produce the less massive
stars that form the peak in the IMF, while accretion processes, which
should operate in a more universal way, may generate the power-law tail
toward high masses.

     Accumulation processes may involve either the coagulation of clumps
(Silk & Takahashi 1979, Pumphrey & Scalo 1982) or the accretion of
diffuse cloud gas by accreting protostellar cores (Zinnecker 1982); in
both cases, power-law mass spectra resembling the observed IMF of
massive stars can be obtained. However, the predictive power of
theories or simulations of these processes is not very great because of
the strong assumptions required. The occurrence of coagulation is in
any case not supported by numerical simulations of cloud collisions
(e.g. Lattanzio et al. 1985), which show that the result is usually
destruction rather than coagulation of the clumps.

     A role for the accretion of diffuse matter in the formation of the
more massive stars is suggested by the simulations of collapse and
fragmentation made by Larson (1978). These simulations showed the
development of hierarchical multiple systems of condensations which
sometimes contain a dominant object that grows by accretion until it has
acquired a significant fraction of the mass in its "accretion domain".
In some cases the results also suggested the development of a

self-similar hierarchy of multiple systems within multiple systems, with
a dominant object or group at each level of the hierarchy. If such
gravitational clustering and accretion processes proceed in a strictly
self-similar way, a power-law mass spectrum would result. Although the
simulations are too crude to predict a mass spectrum with any
confidence, the value of x estimated by counting the objects in the two
largest mass bins is about 1.5, not inconsistent with observations; this
suggests that self-similar gravitational clustering may play a role in
establishing the upper part of the IMF.

Fragmentation

We now turn to fragmentation processes, which can be treated more
quantitatively and which may determine the mass at which the IMF peaks.
The classical fragmentation scheme of Jeans and Hoyle supposes that
small density fluctuations in an initially nearly uniform cloud can grow
as the cloud collapses, and that regions whose mass exceeds a critical
value, the Jeans mass, eventually separate from the rest of the cloud
and collapse on themselves. Since the Jeans mass decreases with
increasing density, fragmentation can in principle continue in a
hierarchical fashion, forming smaller and smaller objects until further
cooling of the gas is finally prevented by increasing opacity. The
smallest fragment mass obtainable in this way is of the order of 0.01 MΘ
(Boss 1986).

This picture may be questioned because the original Jeans analysis
of gravitational instability was mathematically inconsistent in that it
neglected the overall collapse of the fragmenting cloud. Numerical
simulations of cloud collapse and fragmentation actually show little
tendency for fragmentation to occur during the initial nearly free-fall
phase of collapse; instead, small density fluctuations tend to be
smoothed out by pressure forces, and only large-amplitude ones can grow
during this phase (Tohline 1982). If either rotation or a magnetic
field is present, a collapsing cloud first tends to flatten along the
rotation axis or along the field lines to form an equilibrium sheet or
disk; the collapse simulations suggest that fragmentation into smaller
clumps does not become important until such a flattened equilibrium
configuration has formed. There are several advantages in supposing
that fragmentation occurs primarily as a result of the gravitational
instability and breakup of an equilibrium configuration such as a disk,
sheet, or filament: (1) Observations suggest that such configurations
may better approximate the typical structure of star forming clouds than
the extended, nearly uniform medium envisioned in the Jeans analysis;
(2) a rigorous stability analysis can be made; and (3) well-defined
characteristic length and mass scales occur that are determined by the
scale height of the equilibrium configuration. This last feature may be
relevant in accounting for a peak or characteristic mass in the IMF.

Gravitational Stability of Sheets and Filaments

    A number of existing and new results for the gravitational
stability of polytropic sheets and filaments have been assembled by
Larson (1985). Results were derived in particular for values of the
adiabatic exponent γ less than unity that are relevant for star forming
clouds, although the results are qualitatively very similar for all
values of γ. The dispersion relation giving the growth rate of unstable
modes as a function of wavenumber is qualitatively similar for both
sheets and filaments: growth occurs only for wavenumbers less than a
critical value k_c, as in the Jeans problem, but unlike the Jeans result
that the growth rate is a maximum for zero wavenumber, the growth rate
in sheets or filaments reaches a maximum for a wavenumber of about
$0.5\,k_c$ and approaches zero as the wavenumber approaches zero and the
wavelength approaches infinity. The decrease of the growth rate to zero
for infinite wavelength is an important qualitative difference from the
Jeans problem, and it means that the fragmentation of sheets or
filaments is not jeopardized by an overall collapse of the medium.
Instead, perturbations with a characteristic range of sizes or
wavenumbers centered around $0.5\,k_c$ grow faster than those with smaller
or larger sizes and become well separated from the background.

    For an isothermal sheet the critical wavenumber is the inverse of
the scale height H, and the corresponding critical wavelength is $2\pi H$;
the most rapidly growing modes have about half this wavenumber and twice
the wavelength. Somewhat smaller critical wavelengths are obtained for
γ < 1. The critical mass M_c corresponding to k_c depends on the surface
density μ and the temperature T of the fragmenting sheet or filament,
and is proportional to T^2/μ. The coefficient of proportionality has
been given for a number of cases by Larson (1985); typically it is about
2.4 if M_c is measured in solar masses and μ is in $M\odot/pc^2$. The numerical
values are very similar for both sheets and filaments, so the critical
mass can be predicted from the observed quantities T and μ without
knowing the actual three-dimensional geometry of a fragmenting cloud,
i.e. without knowing whether it is filamentary ("spaghetti") or
sheetlike ("lasagna"). In terms of the temperature and the volume
density at the center of a sheet or filament, the critical mass is
essentially the same as the Jeans mass that would be calculated from
these quantities; the essential difference is that the stability results
discussed here have been derived for equilibrium configurations.

    The gravitational stability of sheets and filaments has also been
studied in the presence of rotation and magnetic fields, and a summary
of the results has been given by Larson (1985). In all cases the
qualitative behavior is similar: rotation and magnetic fields both
reduce the growth rates of instabilities, but as long as the instability
is not completely suppressed, the characteristic length and mass scales
are little changed and are still determined by the scale height of the
sheet or filament. With differential rotation, there is an intermediate
regime of incipient instability or "swing amplification" (Toomre 1981)
in which shearing density fluctuations experience finite amplification
as they wind up into trailing spiral features. Associated with such

trailing density enhancements are gravitational torques that tend to
transfer angular momentum outward (Larson 1984) and cause material to
spiral into a single central object, yielding a high efficiency of
condensation and inhibiting further fragmentation.

Comparison with Observations

     In the well-studied nearby Taurus dark clouds the temperature
becomes as low as $\sim$ 8 K and the column density is typically $\sim$ 160
M$\odot$/pc^2, corresponding to a visual extinction AV $\sim$ 8 mag; for these
values the calculated critical mass is about 1 M$\odot$, very similar to the
typical masses of both the dense cloud cores and the T Tauri stars
present (Myers 1986). In larger molecular clouds both T and AV are
higher than in the Taurus clouds; the range of variation of both
quantities is about an order of magnitude, but since the critical mass
is proportional to T^2/AV, the temperature effect predominates and M$_c$ is
higher in the larger and warmer clouds. For example, in clouds with
reflection nebulae a typical temperature is 24 K and the predicted
critical mass is about 6 M$\odot$, while in large molecular clouds with HII
regions the typical temperature is 38 K and the corresponding critical
mass is about 12 M$\odot$ (Larson 1985). These results suggest that the gas
temperature is the dominant parameter controlling the mass scale for
fragmentation.

     It is not yet clear whether the IMF is different in large warm
clouds like the Orion cloud than in small cold clouds like those in
Taurus, but some of the data suggest that this may be the case. The
Orion region of star formation contains seven O stars (Blaauw 1964) and
Taurus contains none, while the numbers of known T Tauri stars in the
two regions are comparable, about 100 in Orion and 75 in Taurus (Cohen &
Kuhi 1979). Moreover, the typical masses of the known T Tauri stars are
somewhat larger in Orion, the median T Tauri mass being about 1.1 M$\odot$ in
Orion and 0.6 M$\odot$ in Taurus (Larson 1982). These differences, if
representative of the total stellar content of the two regions, are
qualitatively in the sense expected from the higher gas temperature in
Orion, which is typically at least 20 K and reaches 75 K near the
Trapezium.

     The relatively high gas temperatures in clouds like the Orion cloud
are almost certainly a result of heating by the most massive and
luminous stars present. The formation of the most massive stars, i.e.
those forming the power-law tail of the IMF, appears to require only a
sufficiently massive molecular cloud, since the mass of the most massive
star in a region of star formation increases systematically with the
mass of the associated molecular cloud (Larson 1982). Thus the
essential requirement for cloud heating and an increased critical mass
is probably just the existence of sufficiently massive clouds condensing
to yield a high local rate of star formation and producing enough very
massive stars to heat the gas. This effect could account qualitatively
for the evidence that typical stellar masses are larger in regions where
the total star formation rate is higher.

If there are two modes of star formation that produce two peaks in the IMF, it is plausible that low-mass star formation occurs in regions like Taurus containing scattered small, cold dark clouds, while the high-mass mode occurs in more active star forming regions such as spiral arms containing massive molecular clouds and giant HII regions. In this context Orion is probably only a mild example of massive star formation, since most of the massive stars in our galaxy and others form in larger complexes of gas and young stars that produce larger HII regions than the Orion Nebula (Elmegreen 1986). In such regions the typical gas temperature is almost certainly higher than in Orion, so the typical stellar mass may well also be higher.

Limits on Fragmentation

Although present data do not definitely establish that star formation is bimodal, or even that the IMF in some regions has a well-defined peak at a mass above one solar mass, the evidence at least seems persuasive that in some regions star formation produces more massive stars and fewer low-mass stars than predicted by a standard Salpeter IMF. This implies that in such regions the continuing fragmentation of a collapsing cloud into objects less massive than a few solar masses must be relatively improbable. Theory and simulations both suggest that the thermal behavior of a collapsing cloud is crucial for the occurrence of fragmentation: fragmentation seems likely to continue as long as the temperature continues to fall with increasing density, but it becomes less likely when the temperature stops decreasing and may stop altogether if the temperature begins to rise again with increasing density (Larson 1985, Boss 1986). Thus the minimum gas temperature attained may play a crucial role in determining the minimum mass of objects that can form by fragmentation.

In the absence of external heating effects, the temperature in a collapsing cloud is predicted to reach a minimum value of about 5 K, which implies a critical mass of about 0.3 M⊙; thus such conditions could plausibly account for low-mass star formation. However, if a strong infrared radiation field is present, as expected in a region of active star formation, the dust in the dense cores of fragmenting molecular clouds is heated to temperatures of the order of 15 - 20 K or more. At densities above about 10^4 hydrogen molecules per cm^3 the gas becomes thermally coupled to the heated dust, causing its temperature to rise with increasing density (Falgarone & Puget 1985); this probably inhibits further fragmentation to smaller masses. For an equilibrium gas sheet, the rise in temperature occurs when the critical mass is about 4 - 6 M⊙; this is illustrated in Figure 1, where the upper two curves, from Falgarone & Puget (1985), show the temperature-density relation for the gas when the dust temperature is 20 K or 15 K, and the diagonal lines are lines of constant M_c. For densities between about 10^4 and 10^5 molecules per cm^3 the critical mass remains nearly constant as the density increases, and this might lead to a peak at this mass in the IMF. The lower panel of Figure 1 shows M_c plotted versus the surface density μ or the visual extinction AV of an equilibrium gas

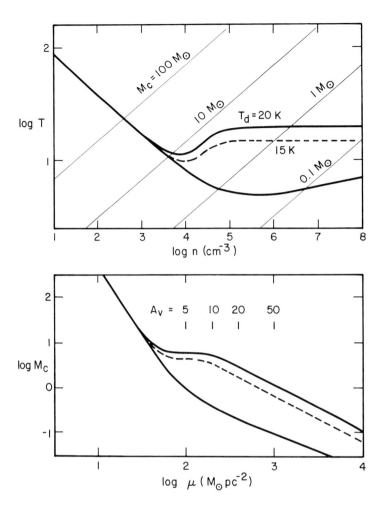

FIGURE 1. The upper panel shows the predicted temperature–density
relation for fragmenting clouds; the lower solid curve (Larson 1985)
applies in the absence of any external heat source, while the upper
solid and dashed curves (Falgarone & Puget 1985) apply when the dust in
cloud cores is heated by external radiation to temperatures of 20 K or
15 K. The diagonal lines indicate the critical mass M_c for an
equilibrium gas sheet as a function of the density and temperature in
the central plane of the sheet. The lower panel gives the critical mass
as a function of the surface density μ of a fragmenting gas sheet or
filament. Some corresponding values of the visual extinction AV are
also indicated; for typical values of AV between 5 and 15 mag, the
critical mass is between 2 and 4 M⊙ when the dust temperature is 15 K
and between 4 and 6 M⊙ when the dust temperature is 20 K.

layer, and for typical values of AV in the range 5 - 15 mag the critical mass is again close to a constant value of several solar masses. Thus circumstances in which the dust in star forming clouds is strongly heated may lead to a high-mass mode of star formation that produces mostly stars more massive than a few solar masses.

Even though the formation of massive stars may be strongly favored when cloud cores are heated, Figure 1 shows that under conditions of exceptionally high density the formation of low-mass stars is still possible. Thus if a large star-forming cloud contains a compact massive core whose density exceeds 10^6 molecules per cm^3, a small dense cluster of low-mass stars may form in this core even though most of the cloud forms massive stars. Such a compact cluster of stars of low or moderate mass is observed in the immediate vicinity of the Trapezium in Orion, and it may survive as a small bound open cluster after the gas and most of the young stars in Orion have been dispersed (Herbig & Terndrup 1986). Thus star clusters may not be very representative of most of the star formation, or of the overall IMF, in the regions in which they formed.

SOME IMPLICATIONS

Interpretations of the observed properties of galaxies have conventionally been based on the assumption of a universal Salpeter-like IMF (e.g. Tinsley 1980), but if this assumption is not valid, many of these interpretations will need to be revised. Unfortunately, since a knowledge of the actual form of the IMF in each system is then crucial, is becomes difficult to make any unique interpretations of the data unless firm constraints can also be placed on the IMF. We have seen that the detailed form of the IMF is still uncertain even in the solar neighborhood; however, as one possible alternative to a Salpeter IMF one can consider a bimodal IMF like that suggested by Larson (1986a) for the solar neighborhood, and ask whether it is any more compatible with the observed properties of galaxies than a Salpeter function.

Integrated Colors and Spectra

If a high-mass mode of star formation is important and the IMF contains a higher proportion of massive stars than a Salpeter function, the predicted colors of galaxies become bluer for a given history of star formation. This could explain why many late-type spiral and irregular galaxies have colors that are too blue (B-V < 0.5, B-H < 3.0) to be explained by conventional models unless relatively young ages are assumed. With a bimodal IMF, it is possible to obtain colors as blue as B-V = 0.2 or B-H = 1.6, comparable to the colors of the bluest normal galaxies, without requiring such young ages (Larson 1986a). A related implication of a bimodal IMF is that the colors and luminosities of starburst galaxies can be explained with less extreme burst parameters; the possibility that part of the color dispersion of peculiar and interacting galaxies could be caused by variations in the IMF was noted by Larson & Tinsley (1978). In general, if the IMF is bimodal or otherwise favors massive stars, the integrated photometric properties of galaxies can be explained with lower present star formation rates than are conventionally required.

Evolution of Galaxies

It has already been noted that the usually derived timescale for gas depletion in spiral and irregular galaxies is uncomfortably short compared to the Hubble time, and that this problem is alleviated if the IMF is bimodal (Sandage 1986). There is little evidence that the gas in spiral and irregular galaxies is replenished by gas infall at the rate required by conventional models. However, if the present SFR in galaxies is smaller than is usually estimated, a smaller rate of gas infall can still be important in helping to sustain the gas content. For example, if star formation is dominated by a high-mass mode, the rate at which gas is permanently locked into remnants can be an order of magnitude or more smaller than the conventionally estimated SFR, so an infall rate an order of magnitude smaller may still have significant effects.

Perhaps the most directly testable prediction of a model in which star formation is bimodal and the formation rate of massive stars declines rapidly with time is that the colors and luminosities of star-forming galaxies should then evolve rapidly with time. Significant changes in galaxy properties with redshift would be predicted even for redshifts as small as 0.5, corresponding to lookback times of the order of 5 billion years. Such effects may have been seen: distant clusters of galaxies have been found to contain a higher proportion of blue, actively star forming galaxies than nearby clusters that are otherwise similar (Butcher & Oemler 1984; Oemler, this conference). More data will be needed to clarify the interpretation of these observations, but the potential of lookback observations for improving our understanding of star formation in galaxies is clear.

Masses and Mass-to-Light Ratios

The most far-reaching implication of models in which the formation rate of massive stars was much higher in the past is probably that a substantial fraction or most of the mass in galaxies is then predicted to be in the remnants of early generations of intermediate- and high-mass stars. In the model of the solar neighborhood that has been discussed, about 55 % of the total mass is in remnants; when similar models are applied to the inner Milky Way and M83, an even larger fractional mass in remnants is implied (Larson 1986a). In starburst galaxies the formation rate of massive stars is so high that stellar remnants must eventually dominate the mass of the starburst region (Weedman 1983). In the nuclear region of M82, for example, the inferred formation rate of massive stars is so high that at this rate all of the

available gas would be converted into stellar remnants in less than a
billion years.

There is every reason to believe that starburst activity was more
prevalent in the past than it is at present. A higher gas content in
galaxies, a greater susceptibility of galactic disks to large-scale
instabilities, and a higher frequency of galaxy interactions and
accretion events would all have led to a higher rate of starburst
activity at earlier times. The higher incidence of blue galaxies in
distant clusters could in part reflect such a higher frequency of
starburst activity in the past. Like the galaxy mergers that apparently
cause many of them, the starbursts that we presently see are probably
"just the dregs of what was once a very common process" (Toomre 1977).
Consequently the ashes of the starburst activity, i.e. stellar remnants,
should now form an important part of the masses of galaxies.

Conventional predictions of galactic mass-to-light ratios (e.g.
Larson & Tinsley 1978) will then require upward revision, as is indeed
necessary to achieve agreement with observed mass-to-light ratios. The
revision should be most important for the galaxies most dominated by
massive stars, i.e. the bluest galaxies, and this will result in a
weaker dependence of mass-to-light ratio on color than is predicted by
conventional models. Such an effect is in fact seen in the data
(Tinsley 1981a), and can be accommodated with a bimodal IMF (Larson
1986a).

Dark Matter as Remnants

What, finally, about the dark matter in galactic halos and clusters
of galaxies? It does not seem very plausible astrophysically that this
dark matter could be in very dim low-mass stars because of the
relatively high temperature of the pre-galactic gas. In contrast, the
possibility that the dark matter is in the remnants of early generations
of massive Population III stars would offer an appealing economy of
hypotheses, since a similar picture can also account for the unseen mass
inside galaxies. If the formation of massive stars was increasingly
favored at earlier times, it is possible that almost exclusively massive
stars were formed at pre-galactic times; a high early star formation
rate, a low abundance of heavy elements, and the relatively high cosmic
background temperature would all have contributed to producing a high
gas temperature and a critical mass probably well above 10 $M\odot$.

If most of the mass in the universe is in the remnants of massive
Population III stars, a very high luminosity must have been produced in
the pre-galactic era; however, it would not conflict with observed
limits on the cosmic background radiation provided that these massive
stars formed within the first billion years of the history of the
universe (Carr, Bond, & Arnett 1984). Nucleosynthesis considerations
also do not at present place any strong contraints on the possible early
formation of large numbers of massive Population III stars, since many
of these stars may simply have collapsed to black holes without

contributing to heavy element production (Larson 1986b).

The most implausible aspect of remnants as a candidate for the dark mass has always seemed to be the extreme requirements placed on the rate and efficiency of pre-galactic massive star formation: 90 % or more of the matter in the universe must in this case have been transformed into the remnants of massive stars in less than a billion years. However, these extreme requirements are similar to what is actually observed, at least on a small scale, in starburst systems like M82, where the rate of massive star formation is sufficient to turn all of the gas into remnants in less than a billion years. If starburst activity dominated the universe in the pre-galactic era, and if almost exclusively massive stars were formed, it seems not impossible that most of the mass in the universe could indeed be in dead stars.

REFERENCES

Armandroff, T. E. (1983). The effect of changes in the luminosity function on problems in galactic structure. The Nearby Stars and the Stellar Luminosity Function, ed. A. G. D. Philip & A. R. Upgren, pp. 229-33. L. Davis Press, Schenectady.
Augarde, R. & Lequeux, J. (1985). Peculiar motions and star formation in the interacting galaxy complex Mk 171 = NGC 3690 + IC 694. Astr. Astrophys., 147, 273-80.
Bahcall, J. N. (1986). Dark matter in the galactic disk. Dark Matter in the Universe, IAU Symp. No. 117, ed. J. Kormendy & G. R. Knapp, in press. Reidel, Dordrecht.
Blaauw, A. (1964). The O associations in the solar neighborhood. Ann. Rev. Astr. Astrophys., 2, 213-46.
Boss, A. P. (1986). Protostellar formation in rotating interstellar clouds. V. Nonisothermal collapse and fragmentation. Astrophys. J., in press.
Butcher, H. & Oemler, A. (1984). The evolution of galaxies in clusters. V. A study of populations since z ∿ 0.5. Astrophys. J., 285, 426-38.
Carlberg, R. G., Dawson, P. C., Hsu, T. and VandenBerg, D. A. (1985). The age-velocity-dispersion relation in the solar neighborhood. Astrophys. J., 294, 674-81.
Carr, B. J., Bond, J. R. & Arnett, W. D. (1984). Cosmological consequences of population III stars. Astrophys. J., 277, 445-69.
Cohen, M. & Kuhi, L. V. (1979). Observational studies of pre-main-sequence evolution. Astrophys. J. Suppl., 41, 743-843.
Elmegreen, B. G. (1986). Large scale star formation: density waves, superassociations and propagation. Star Forming Regions, IAU Symp. No. 115, ed. M. Peimbert & J. Jugaku, in press. Reidel, Dordrecht.
Falgarone, E. & Puget, J. L. (1985). A model of clumped molecular clouds. I. Hydrostatic structure of dense cores. Astr. Astrophys., 142, 157-70.

Freedman, W. L. (1985). The upper end of the stellar luminosity
     function for a sample of nearby resolved late-type galaxies.
     Astrophys. J., 299, 74-84.
Freeman, K. C. (1977). Star formation and the gas content of galaxies.
     The Evolution of Galaxies and Stellar Populations, ed. B. M.
     Tinsley & R. B. Larson, pp. 133-56. Yale University Observatory,
     New Haven.
Gusten, R. & Mezger, P. G. (1983). Star formation and abundance
     gradients in the Galaxy. Vistas Astr., 26, 159-224.
Herbig, G. H. & Terndrup, D. M. (1986). The Trapezium cluster of the
     Orion nebula. Astrophys. J., 307, in press.
Jensen, E. B., Talbot, R. J. & Dufour, R. J. (1981). M83. III. Age
     and brightness of young and old stellar populations. Astrophys.
     J., 243, 716-35.
Jones, T. J., Hyland, A. R., Straw, S., Harvey, P. J., Wilking, B. A.,
     Joy, M., Gatley, I. & Thomas, J. A. (1986). Star formation in the
     Magellanic Clouds - III. IR observations of giant HII regions.
     Mon. Not. Roy. Astr. Soc., 219, 603-27.
Kennicutt, R. C. (1983). The rate of star formation in normal disk
     galaxies. Astrophys. J., 272, 54-67.
Kronberg, P. P., Biermann, P. & Schwab, F. R. (1985). The nucleus of
     M82 at radio and X-ray bands: discovery of a new population of
     supernova candidates. Astrophys. J., 291, 693-707.
Lacey, C. G. & Fall, S. M. (1985). Chemical evolution of the galactic
     disk with radial gas flows. Astrophys. J., 290, 154-70.
Larson, R. B. (1977). Rates of star formation. The Evolution of
     Galaxies and Stellar Populations, ed. B. M. Tinsley & R. B. Larson,
     pp. 97-132. Yale University Observatory, New Haven.
Larson, R. B. (1978). Calculations of three-dimensional collapse and
     fragmentation. Mon. Not. Roy. Astr. Soc., 184, 69-85.
Larson, R. B. (1982). Mass spectra of young stars. Mon. Not. Roy.
     Astr. Soc., 200, 159-74.
Larson, R. B. (1984). Gravitational torques and star formation. Mon.
     Not. Roy. Astr. Soc., 206, 197-207.
Larson, R. B. (1985). Cloud fragmentation and stellar masses. Mon.
     Not. Roy. Astr. Soc., 214, 379-98.
Larson, R. B. (1986a). Bimodal star formation and remnant-dominated
     galactic models. Mon. Not. Roy. Astr. Soc., 218, 409-28.
Larson, R. B. (1986b). Dark matter: dead stars? Comments Astrophys.,
     11, in press.
Larson, R. B. & Tinsley, B. M. (1978). Star formation rates in normal
     and peculiar galaxies. Astrophys. J., 219, 46-59.
Larson, R. B., Tinsley, B. M. & Caldwell, C. N. (1980). The evolution
     of disk galaxies and the origin of S0 galaxies. Astrophys. J.,
     237, 692-707.
Lattanzio, J. C., Monaghan, J. J., Pongracic, H. & Schwarz, M. P.
     (1985). Interstellar cloud collisions. Mon. Not. Roy. Astr. Soc.,
     215, 125-47.
Mezger, P. G. & Smith, L. F. (1977). Radio observations related to star
     formation. Star Formation, IAU Symp. No. 75, ed. T. de Jong & A.
     Maeder, pp. 133-63. Reidel, Dordrecht.
Miller, G. E. & Scalo, J. M. (1979). The initial mass function and

stellar birthrate in the solar neighborhood. Astrophys. J. Suppl.,
    41, 513–47.
Mirabel, I. F. & Morras, R. (1984). Evidence for high–velocity inflow
    of neutral hydrogen toward the Galaxy. Astrophys. J., 279, 86–92.
Myers, P. C. (1986). Dense cores and young stars in dark clouds. Star
    Forming Regions, IAU Symp. No. 115, ed. M. Peimbert & J. Jugaku, in
    press. Reidel, Dordrecht.
Oort, J. H. (1965). Stellar dynamics. Galactic Structure, ed. A.
    Blaauw & M. Schmidt, pp. 455–511. University of Chicago Press.
Philip, A. G. D. & Upgren, A. R. (1983). The Nearby Stars and the
    Stellar Luminosity Function (editors), IAU Colloq. No. 76. L.
    Davis Press, Schenectady.
Pumphrey, W. A. & Scalo, J. M. (1983). Simulation models for the
    evolution of cloud systems. I. Introduction and preliminary
    simulations. Astrophys. J., 269, 531–59.
Quirk, W. J. & Tinsley, B. M. (1973). Star formation and evolution in
    spiral galaxies. Astrophys. J., 179, 69–83.
Rieke, G. H., Lebofsky, M. J., Thompson, R. I., Low, F. J. & Tokunaga,
    A. T. (1980). The nature of the nuclear sources in M82 and NGC
    253. Astrophys. J., 238, 24–40.
Salpeter, E. E. (1955). The luminosity function and stellar evolution.
    Astrophys. J., 121, 161–7.
Sandage, A. (1986). Star formation rates, galaxy morphology, and the
    Hubble sequence. Astr. Astrophys., in press.
Scalo, J. M. (1986). The stellar initial mass function. Fundam. Cosmic
    Phys., 11, 1–278.
Schmidt, M. (1963). The rate of star formation. II. The rate of
    formation of stars of different mass. Astrophys. J., 137, 758–69.
Silk, J. & Takahashi, T. (1979). A statistical model for the initial
    stellar mass function. Astrophys. J., 229, 242–56.
Tinsley, B. M. (1980). Evolution of the stars and gas in galaxies.
    Fundam. Cosmic Phys., 5, 287–388.
Tinsley, B. M. (1981a). Correlation of the dark mass in galaxies with
    Hubble type. Mon. Not. Roy. Astr. Soc., 194, 63–75.
Tinsley, B. M. (1981b). Chemical evolution in the solar neighborhood.
    IV. Some revised general equations and a specific model.
    Astrophys. J., 250, 758–68.
Tohline, J. E. (1982). Hydrodynamic collapse. Fundam. Cosmic Phys., 8,
    1–82.
Toomre, A. (1977). Mergers and some consequences. The Evolution of
    Galaxies and Stellar Populations, ed. B. M. Tinsley & R. B. Larson,
    pp. 401–26. Yale University Observatory, New Haven.
Toomre, A. (1981). What amplifies the spirals? The Structure and
    Evolution of Normal Galaxies, ed. S. M. Fall & D. Lynden–Bell, pp.
    111–36. Cambridge University Press.
Twarog, B. A. (1980). The chemical evolution of the solar neighborhood.
    II. The age–metallicity relation and the history of star formation
    in the galactic disk. Astrophys. J., 242, 242–59.
van den Bergh, S. (1972). Stellar populations in galaxies. External
    Galaxies and Quasi–Stellar Objects, IAU Symp. No. 44, ed. D. S.
    Evans, pp. 1–11. Reidel, Dordrecht.
Viallefond, F. & Thuan, T. X. (1984). A multifrequency study of star

formation in the blue compact dwarf galaxy I Zw 36. Astrophys. J.,
    269, 444–65.
Weedman, D. W. (1983). Toward explaining Seyfert galaxies. Astrophys.
    J., 266, 479–84.
Zinnecker, H. (1982). Prediction of the protostellar mass spectrum in
    the Orion near-infrared cluster. Symposium on the Orion Nebula to
    Honor Henry Draper, ed. A. E. Glassgold, P. J. Huggins & E. L.
    Schucking, pp. 226–35. Ann. New York Acad. Sci., 395.

DISCUSSION

LEQUEUX: It seems that your model is producing far too much iron. If you can drop the oxygen production by introducing a cutoff at $\sim$ 17 M$\odot$ above which stars do not explode, this trick cannot solve the problem of iron and other elements like carbon which are supposed to be produced in lower-mass stars. What can you do in this respect?

LARSON: I didn't address the question of iron production because it has been my impression that the source and production rate of iron are too uncertain for quantitative predictions to be made with any confidence. I hope those with more expertise in this area will pursue this question further.

WYSE: Joe Silk and I have been working on models with a bimodal IMF which are somewhat different from Richard Larson's, and we specifically attempt to fit the age-metallicity relation for the solar neighborhood, with an acceptable gas consumption time etc. The models provide a reasonable fit to the data.

G. BURBIDGE: Following up on the question of enrichment it is obvious that an early epoch in which massive stars predominate will lead to the production and ejection of helium and heavier elements, and such a phase will have an effect on the so-called cosmological helium abundance.

   As far as individual galaxies are concerned, I have done some work on the very luminous (> 10^{12} L$\odot$) IRAS galaxies. For cases where the effect is not due to activity in the nucleus I have concluded that they may be genuinely young galaxies comprising a few hundred thousand stars with masses in the range 20 -120 M$\odot$. They not only produced the light but also the elements which condense into the dust, leading to dust shells which give luminosity in the 60 - 100 μ region. The prototype star which does just this in our own Galaxy is Eta Carinae. It is making dust by unknown processes at a rate of about 10^{-2} M$\odot$/year. Thus very bright IRAS galaxies may be young. The time scale associated with the phenomenon is $\sim 10^{8}$ years, involving several generations of $\sim$ 100 M$\odot$ stars probably forming in several small clusters.

LARSON: Concerning helium production in my model of galactic evolution, the estimated helium production satisfies $\Delta Y/\Delta Z \sim 2 - 3$ and therefore does not conflict with observed limits on helium abundance variations. Pre-galactic production of helium by Population III stars was considered by Carr, Bond & Arnett (1984), and they showed that "very massive objects" could produce a lot of helium without overproducing heavier elements; in fact it is possible that the entire "primordial" helium abundance could be produced this way.

   I agree that the picture of superluminous IRAS galaxies as starburst systems, possibly very young, dominated by massive stars is an appealing one. They are probably the most extreme examples of starburst galaxies, and M82, which I discussed, may just be a much more modest nearby example.

RENZINI: ZAMS models indicate a <u>sharp</u> change in the slope of the
mass-luminosity relation around 0.6 M0, with the exponent decreasing
from $\sim$ 5 to $\sim$ 2.5. The derived IMF around 0.6 M0 is therefore crucially
dependent on this feature, and its possible metallicity dependence.

LARSON: I agree that the IMF is quite sensitive to the form of the
mass-luminosity relation (MLR) and that one needs to consider carefully
possible uncertainties or variations in the MLR. Scalo (1986) discusses
this subject at some length, including both the empirical and the
theoretical MLR, and adopts a curve that gives a good representation of
all of the available data. There is no indication in either the data or
the model results of variations in the slope of the MLR large enough to
alter the qualitative form of the derived IMF, e.g. to remove the dip at
0.7 M0, although such variations probably can't be excluded as outside
the possible uncertainties.

KING: A limit on the masses of remnants is set by the recent
demonstration that if the unseen mass in the solar neighborhood is in
chunks larger than 2 solar masses, they would break up the wide binaries
(Bahcall, Hut & Tremaine, Astrophys. J., <u>290</u>, 15, 1985).

LARSON: This constraint is easily satisfied in a remnant-dominated
model; in my model of the solar neighborhood most of the unseen mass is
in white dwarfs with masses less than 1.4 M0. A more troublesome
constraint is that a model should not predict too many observable white
dwarfs; even a standard model (constant SFR, conventional IMF) has
difficulty satisfying this constraint. It seems possible that most of
the white dwarfs making up the unseen mass could be relatively massive,
$\sim$ 1 M0 or more, in which case they are predicted to fade to invisibility
in less than 10 Gyr.

ZINNECKER: I have a few comments:
(1) The Salpeter slope would not be x = 1.35 but x = 1.05 if Salpeter
in 1955 had done the analysis under the same assumptions as used in
later determinations of the IMF (Taff, Astr. J., <u>79</u>, 1280, 1974). So
the IMF in star clusters is perhaps significantly steeper than that of
the field stars in the 1 - 10 M0 range.
(2) Even if the observed field star luminosity function flattens and
turns over at faint magnitudes, the corresponding IMF need not have a
turnover. This is because the transformation from the LF to the IMF
involves the <u>derivative</u> of the mass-luminosity relation and this
relation may not be a straight line, either observationally or
theoretically (D'Antona & Mazzitelli, Astr. Astrophys., <u>127</u>, 149, 1983).
(3) Don't you have a problem with the white dwarf remnants that are
predicted in your bimodal galactic evolution model? If white dwarfs are
to make up most of Bahcall's unaccounted mass in the solar neighborhood,
why would these white dwarfs be unobservably faint?

LARSON: (1) Scalo (1986) has analyzed all of the available data for
both field stars and open clusters with the same assumptions and does

not find that the cluster IMF is steeper than the field star IMF; in fact he concludes that the open cluster IMF may, if anything, be <u>flatter</u> than the field star IMF for the largest masses.
(2)′ Again I refer to Scalo's (1986) review where this question is discussed. Even though the mass-luminosity relation is uncertain for the faintest stars, it doesn't appear that any variation in its slope allowed by the data could yield an IMF that rises at the lowest observed masses. However, a steep drop at small masses should probably not be considered as well established.
(3) As I suggested in my answer to King, most of the white dwarfs might be massive enough (> 1 M$\odot$) to fade to invisibility in less than 10 Gyr. In any case, even a standard model predicts too many observable white dwarfs, so it is possible that there is a more fundamental problem with white dwarf theory that needs to be resolved before conclusions about galactic evolution and the IMF can be drawn.

DRESSLER: The strength of the bimodality in your model for the solar neighborhood depends on an assumed age of the population of 15 Gyr. The only evidence you presented for this was the age of the disk globulars, however these are representative of the <u>inner</u> disk, not the solar neighborhood. What other evidence from stellar populations supports this assumption that the disk at the location of the Sun is, in fact, older than 10 Gyr?

LARSON: None. However, it has been my impression that Zinn's disk population of globular clusters extends out as far as the Sun's distance from the galactic center.

JANES: Regarding the question of the age of the galactic disk, the limited evidence that exists suggests that it is substantially younger than 15 Gyr. The oldest open clusters are less than 10 billion years old (but they are also easily disrupted). The lower envelope of the field color magnitude diagram matches NGC 188, a cluster of some 8 - 10 billion years in age (in this case, metallicity effects could give the same result). Finally, Twarog's field star samples include a rather small proportion of very old stars.

LARSON: If the disk is younger than 15 Gyr the size of the bump at $\sim$ 1 M$\odot$ in the derived IMF is reduced, but to make it insignificant you have to make the disk younger than 10 Gyr <u>and</u> require a non-decreasing SFR. In any case, the existence of a disk population of globular clusters shows that <u>some</u> star formation was occurring in the disk 15 Gyr ago. Maybe predominantly massive stars formed before 10 Gyr ago, and low-mass star formation didn't get under way in the local galactic disk until $\sim$ 10 Gyr ago.

ROMAN: What is the effect on the derived IMF of a recent (compared to the age of the galaxy) burst of formation of high mass stars?

LARSON: If the massive stars in the local galactic disk were formed by a recent burst of star formation, the derived IMF for massive stars would be overestimated if a constant or decreasing SFR were assumed.

Scalo (1986) considered the possibility of an increasing SFR, and in this case the size of a possible second peak is greatly reduced. The dip at 0.7 M☉ and the upturn between 0.7 and 0.85 M☉ are unaffected by assumptions about the age of the disk or the time dependence of the SFR, but I wouldn't want to stake my career on this evidence for bimodality of the IMF.

KOO: Given the very old ages, low metallicities, and low M/L of the oldest globular clusters, does your model of bimodal star formation require that such agglomerations formed only in very cool clouds? What about our halo?

LARSON: In the written version of my talk I suggest that bound star clusters may form in the exceptionally dense cores of some massive star forming clouds where the high density leads to a small critical mass despite the relatively high ambient temperature. Most of the cloud may form more massive stars in a looser aggregation that does not survive as a bound cluster. Scaled up a few orders of magnitude, what we see in Orion may provide a model for the formation of globular clusters. In such a picture, most of the star formation in the halo produced massive stars, and globular clusters are exceptional objects that formed in conditions of exceptionally high density.

SCHOMMER: Would you expect a significant population of 1 M☉ white dwarfs in globular clusters? If they cool rapidly and are the dominant population of white dwarfs, they may be very hard even for ST to detect, or at least the detection may depend quite sensitively on the mass function. Gary Da Costa reminds me that dynamical models of globular clusters, in particular 47 Tuc, require a small, but significant population of $\sim$ 1.2 M☉ collapsed objects.

LARSON: Quite possibly you would expect globular clusters to contain relatively massive white dwarfs. Apart from the possible contribution of a high-mass mode of star formation, which may have been important generally but may not have contributed much to the population that survives in a bound cluster (see above), stellar remnants are probably more massive in metal-poor populations because the expected mass loss rate is lower for metal-poor stars. (Such an effect might also make the remnants of early disk stars more massive and help to solve the problem that fewer faint white dwarfs are seen than are predicted.)

STAR FORMATION HISTORIES OF GALACTIC DISKS

R.C. Kennicutt
Department of Astronomy, University of Minnesota
116 Church Street, SE, Minneapolis, MN 55455

Abstract. The observational constraints on stellar birthrate
histories in galactic disks are reviewed. Observations of
individual stars in the Galactic disk, the broadband optical
colors of galaxies, and the star formation rates in the
Galaxy and in external galaxies are all consistent with a
common picture in which most of the photometric properties
of disks are due to differences in their star formation
histories. Possible sources of systematic error in the
observations are also discussed. The inferred birthrate
histories are much flatter with time than is expected from
simple closed-system galaxy evolution models. Several
possible modifications to the models are discussed, including
bimodal mass functions, infall and radial inflows, and modifications to the basic star formation law.

INTRODUCTION

Understanding the star formation histories of galaxies is one of
the most important and challenging problems in the field of galaxy
evolution. Even relatively crude observational constraints on birthrate histories enable us to test the basic assumptions which underlie
galactic and chemical evolution models, and provide necessary input
data for other problems such as the stellar initial mass function (IMF).
Actually measuring the star formation history of a galaxy is extremely
difficult, however, and current observations provide only rudimentary
constraints on the birthrate history of even the Galactic disk.
Nevertheless, the available data provide a surprisingly self-consistent
observational picture of the evolution of galactic disks, and one which
has important consequences for the theoretical understanding of galaxy
evolution.

In this paper I shall review the current observational constraints
on the birthrate histories of galactic disks, and discuss the results
in the context of simple evolutionary models. Although the basic
picture of disk evolution provided by observations has not changed
appreciably over the past decade, several recent studies of star
formation both in the Galaxy and in external galaxies have put these
results on a much firmer footing. Taken together, the observations

suggest that galactic disks have evolved at a relatively steady rate over their lifetimes, a result which is in fundamental disagreement with simple closed models of galactic evolution. Again this discrepancy is nothing new; it was identified by Schmidt over 20 years ago, but it has yet to be resolved to anyone's satisfaction. On the theoretical side, the apparent inconsistencies between the observations and the traditional evolution models have led several investigators to propose alternative mechanisms for regulating the birthrates or the gas contents of galaxies, and I shall briefly review this work in the last section. This has been a very active field recently, both observationally and theoretically, and one which will probably mature rapidly over the next few years.

During the preparation of this paper I have benefited greatly from the earlier reviews by Tinsley (1980), Kron (1982), Güsten and Mezger (1983), and Lequeux (1985), and the reader is referred to these papers for more general discussions. I have chosen to restrict this discussion to disk evolution; the evolution of elliptical and irregular galaxies are reviewed elsewhere in this volume by O'Connell, Aaronson, and Sargent.

OBSERVATIONAL CONSTRAINTS ON BIRTHRATE HISTORIES OF GALACTIC DISKS

Stellar Statistics

Ideally, one would like to determine the star formation history of a galaxy directly, by extracting a representative region, isolating all of the stars in a given mass range, and measuring the ages of the stars one by one. This is obviously unrealistic for external galaxies, but one can perform at least an approximation to this exercise in the solar neighborhood, and it is such measurements which provide the most fundamental information on birthrate histories.

Twarog (1980) has studied the birthrate history of F dwarfs in the solar neighborhood, by obtaining Strömgren photometry for an unbiased sample of stars, and using theoretical isochrones to estimate ages for each star. The derived star formation rate (SFR) history for the disk is shown in Figure 1. Three relations are shown, corresponding to three different assumed slopes for the IMF over the relevant mass range. Consideration of several selection effects in the derivations, as well as comparison with the independently derived age-metallicity relation for the same stars, suggest that the lowest values in Figure 1 are the most likely.

Twarog's results indicate that the SFR in the Galactic disk, at least for solar-mass stars, has been roughly constant, never rising by more than 2-3 times the present rate over the history of the disk. A convenient parameter for subsequent comparisons is the ratio of the present SFR to the average SFR in the past. Following the notation of Scalo (1986):

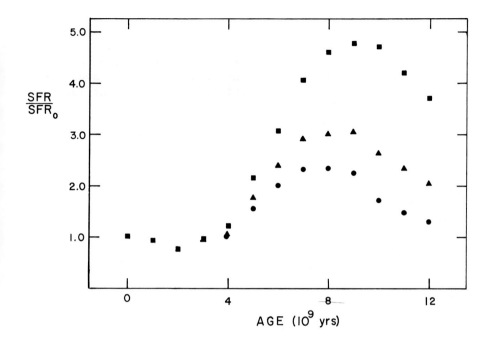

$$\frac{SFR}{SFR_0}$$

AGE (10^9 yrs)

Figure 1.   Star formation history for the local Galactic disk, from Twarog (1980).   SFR is normalized to the present rate, and zero age corresponds to the present time.   Symbols correspond to different IMF slopes.

$$b \equiv \frac{R(present)}{<R(t)>} \tag{1}$$

Twarog derives a firm lower limit b > 0.4 for the solar neighborhood, with the most probable values being between 0.6 and 1.0.   It is important to note that Twarog's measurements apply only to relatively low-mass stars.   Nevertheless his results indicate that the dominant component of the stellar population in the Galactic disk has formed at nearly a constant rate since the disk's formation.

Another, less direct argument for a relatively constant birthrate in the Galactic disk has been made by Tinsley (1977), Miller and Scalo (1979), and Scalo (1986), based on the shape of the present-day mass function (PDMF) in the solar neighborhood.   Stars in the 1-2 $M_\odot$ range possess lifetimes which are comparable to the age of the disk, so the determination of the time-averaged IMF in this mass range depends on both the observed PDMF and the birthrate history of these stars.   If we assume a priori that the true IMF is continuous over the 1-2 $M_\odot$ range, then this continuity constraint can be used to place rough limits on the birthrate history.   Applying this constraint to the solar neighborhood data constrains the relative birthrate b to the range 0.2<b<2.5, roughly consistent with Twarog's range of values (Scalo 1986).   The reader is referred to Scalo's review article for a much more thorough discussion of this method and its limitations.

This IMF continuity argument is of course not a direct measurement of the birthrate history of the disk, but is merely a consistency argument. The same argument can be turned around and used to content that if the IMF is really discontinuous near 1 $M_\odot$, the stellar birthrate must have changed dramatically over the history of the disk. Larson (1986) uses precisely such an argument to support his bimodal IMF model (also see Larson's review in this volume), and hence the shape of the PDMF by itself cannot provide unambiguous information on the stellar birthrate history. As emphasized by Scalo (1986), however, it provides an interesting point of comparison with the other determinations of the star formation histories.

The only other galaxy for which we have even crude stellar data relevant to this problem is the Large Magellanic Cloud. Several studies of the field star luminosity function in the LMC have been published (e.g., Butcher 1977, Stryker and Butcher 1981, Hardy et al. 1984), and the data have been used to approximately date the major star forming episodes in the galaxy. Those workers conclude that the bulk of the star formation in the LMC has occurred over time scales which are recent relative to the total age of the galaxy.

Models of Galaxy Colors

For external galaxies we cannot obtain data on the ages of individual stars, so everything that we know about their star formation histories comes from interpretation of their integrated light. The first and most widely applied method is the modelling of the optical broadband colors of galaxies.

The basic model building procedures have been described in detail in the early studies by Tinsley (1968, 1972), Searle et al. (1973), and Larson and Tinsley (1978), and most of the more recent applications, too numerous to cite individually here, have followed essentially the same approach. Evolutionary tracks (converted to observable luminosities and colors using model stellar atmospheres) for stars of discrete masses, or isochrones for populations of discrete ages, are combined on the computer to synthesize the luminosities and colors of populations with different choices for the IMF, composition, and age. The effect of varying the stellar birthrate can be investigated by varying the age mix of stars. The birthrate history used in the models is usually parametrized as an exponential function of time:

$$R(t) = R_o \, e^{-\beta t} \qquad (2)$$

As will be shown below, the choice of the precise functional form for R(t) is not critical, because the galaxy colors are only sensitive to the rough time dependence of the SFR. Models of this sort, though they may appear to be a programmer's nightmare at first glance, are actually quite straightforward to program on even a small modern computer, and it is quite easy to generate vast piles of galaxy models with a "numerical accuracy far exceeding the astrophysical certainty of the calculation" (Tinsley 1980).

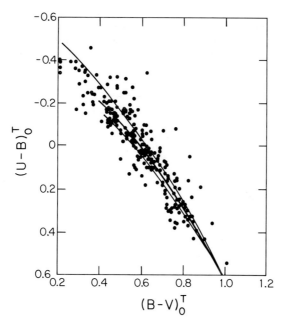

Figure 2.   Observed (corrected) RC2 colors for spiral galaxies. Superimposed are 3 evolutionary models, with constant IMF and variable birthrate history.

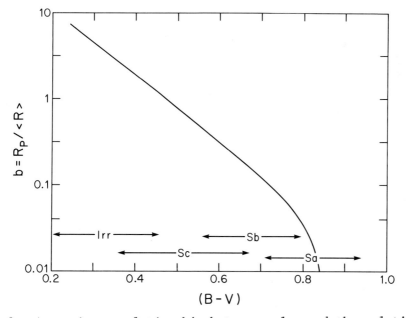

Figure 3.   Approximate relationship between color and the relative SFR parameter b, for the best fitting model.   Approximate observed color ranges for different galaxy types are also shown.

The results from one such calculation are illustrated in Figure 2, taken from Kennicutt (1983). Shown are the reddening-corrected UBV colors of spiral galaxies from the Second Reference Catalog of Bright Galaxies (de Vaucouleurs et al. 1976), along with three sets of evolutionary models. Note that while the galaxies exhibit a considerable scatter in colors, they follow a relatively well defined sequence which is well represented by the models. Each line in Figure 2 is a set of pure-disk models, in which the age, composition, and IMF are held fixed, and only the star formation history is varied. The three curves show different choices for the IMF, with the middle curve corresponding to roughly a Salpeter function. This set of models provides a good fit to the run of observed galaxy colors, and demonstrates that the UBV colors of normal galaxies can be reproduced with a one-parameter sequence of models, with star formation history the dominant variable (Tinsley 1968, Searle et al. 1973).

The models can then be used to infer the range of birthrate histories which is needed to account for the observed galaxy colors. Figure 3 shows the relationship between color and the birthrate parameter b and the e-folding rate β for a model with the best fitting IMF, solar composition, and a total disk age of 15 Gyr. The model B-V colors range from about 0.50 for a galaxy with a uniform SFR with time ($\beta = 0$, $b = 1$), to roughly 1.0 for a single-burst old disk ($\beta = \infty$, $b = 0$). Also shown in Figure 3 are the approximate ranges in observed colors for different galaxy types. Hence the colors of typical Sc galaxies imply roughly constant birthrates in these systems, while in earlier type spirals the birthrate has decreased with time, although not dramatically. Also note that galaxies which are bluer than B-V $\cong$ 0.50 can only be explained by a present SFR which is higher than the average past rate, if one assumes that their other properties (e.g., IMF) are the same as for the redder galaxies.

How accurately can one determine the SFR history of a galaxy from its integrated colors? Unfortunately the systematic uncertainty in the birthrate-color relation is considerably larger than would be suggested by the apparently excellent agreement in Figure 2. The principal reason is that integrated colors of disk galaxies are overwhelmingly dominated by two relatively narrow regions of the H-R diagram, and the relative contributions of these two stellar components are only sensitive not only to the SFR history but also to a variety of other properties of the stellar populations.

An excellent illustration of the dominant contributors to the integrated light of a typical (our own!) galactic disk is shown in Figure 4 (taken from Kron 1982), which displays the frequency of stellar types recorded in the first three volumes of the Michigan Spectral Catalog (Houk 1982). The distribution of stars in this diagram provides a realistic indication of the stars which would dominate the integrated spectrum of the galactic disk if it were observed from the outside (the obvious incompleteness for faint stars does not affect the arguments here). Most of the disk light is

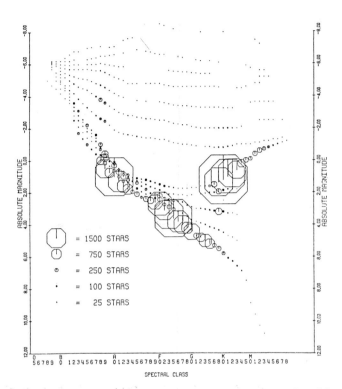

HR DIAGRAM
VOLUMES 1, 2, AND 3

Figure 4.   Frequency of stellar spectral types from the first 3 volumes
of the Michigan Spectral Catalogue.   Figure reproduced courtesy of
N. Houk.

contributed by A-type dwarfs, with ages of order 10^7-10^8 yr, and K-type
giants, with ages of 10^8-10^{10} yr.   Hence the integrated UBV colors
measure not the detailed star formation history but rather the ratio of
the current SFR (for intermediate-mass stars) to the average past rate
(for roughly solar-mass stars), essentially the birthrate parameter b
defined earlier.   More detailed information about the birthrate history
cannot be derived from the UBV colors alone.   Although other colors,
such as the near-infrared V-K index, are sensitive to slightly different
stellar components, their sensitivity to the detailed SFR history is no
better (e.g., Aaronson 1978).

     As a result of this segregation of the dominant contributors to
the disk light, the integrated colors are sensitive to a number of
other properties of the stellar population, including (in roughly
decreasing order of importance), the IMF, uncertainties in post giant
branch evolution, reddening, contamination by bulge light, metallicity,

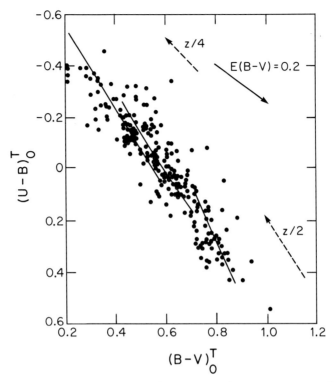

Figure 5.   Same data as in Figure 2.   Here the curves are for constant-
birthrate history models, wich b = 0, .15, and .07.   Also shown are the
effects of decreasing the metallicity of the model galaxies, and the
interstellar reddening line.   The metallicity calculations are from
Larson and Tinsley (1978).

and disk age.   Systematic changes in these secondary parameters usually
alter the colors in ways which closely mimic changes in birthrate
history.   This is illustrated in Figure 5, which shows the same data as
in Figure 2, but in this case shown with 3 sequences in which the
birthrate history is fixed, and the IMF is varied.   Figure 2 and 5
demonstrate that the effects of varying the birthrate history and the
IMF are nearly degenerate in the two-color diagram, and hence we cannot
exclude the possibility of substantial IMF variations along the Hubble
sequence, along with very different star formation histories than are
implied by the "standard" fixed-IMF models.   Figure 5 also illustrates
the effects of varying metallicity and reddening, as adapted from
Larson and Tinsley (1978); these are also virtually degenerate with
the birthrate sequence, as are the other parameters mentioned earlier.

My intention in raising these caveats is not to question the
general picture of star formation histories revealed by the color
models, but simply to discourage the reader from overinterpreting
diagrams such as Figure 3.   For individual galaxies, uncertainties in
IMF, reddening, etc., probably render any determination of the star

formation history uncertain at the factor-of-two level at best for blue galaxies, and much worse in redder galaxies, where the dependence of color on birthrate is weak. It is also possible that systematic changes in the IMF, etc., along the Hubble sequence affect the overall birthrate scale which is derived from the simple one-parameter models, but the consistency between the color models and other birthrate indicators, including the solar neighborhood studies discussed earlier, suggests that the Searle et al. (1973) birthrate picture is still basically correct. I shall return to this point at the end of the section.

Measurements of Current Star Formation Rates

Over the past few years, several new observational techniques for estimating the current SFRs in galaxies have been developed, and these data can also be used to constrain the star formation histories. The average past rate of star formation in a disk is simply its stellar mass divided by its age; this can be estimated quite easily from photometry, and hence if one also has a means of estimating the current SFR, the birthrate parameter b can be determined directly. Although a detailed review of the work on SFRs is beyond the scope of this paper, a brief summary of the current observational methods is useful.

Ultraviolet Emission: UV emission in the 1000-3000Å range directly traces the young stellar component. Surveys of the star formation in nearby galaxies have been published by Vangioni-Flam et al. 1980, Rocca-Volmerange et al. (1981), and Donas and Deharveng (1984). The biggest limitations of this method are the severe extinction at these wavelengths, and the current paucity of good data, though this situation will change dramatically with the launching of the HST and ASTRO observatories.

H-Alpha Emission: The Balmer emission of a galaxy provides an easily measured, direct tracer of its massive (>10 $M_\odot$) SFR, and has been applied extensively by Kennicutt (1983), Kennicutt and Kent (1983), and Gallagher et al. (1984). The results of this work will be discussed in more detail below.

Free-Free Radio Emission: Another variant of the Lyman photon-counting method is to measure the thermal radio luminosity. The primary weakness of this method, as with H-alpha, is the limitation to the upper end of the stellar IMF, but it avoids the problems with extinction which affect the optical and UV data. This method has been employed effectively in the Galaxy by Smith et al. (1978) and by Güsten and Mezger (1983), and provides the best data by far on the global SFR in the Galaxy. Studies of external galaxies are hampered by the overwhelming contamination by nonthermal radio emission at most wavelengths (e.g., Gioia et al. 1982, Israel and van der Hulst 1984), and have been limited to a few nearby systems (e.g., Klein et al. 1984).

Far Infrared Emission: This is potentially a very powerful star formation tracer (e.g., Rieke et al. 1980, Gehrz et al. 1983, Hunter et al. 1986, Thronson and Telesco 1986). The quantitative determination of SFRs in normal galaxies has proven to be problematical, however. Modelling of the emission has revealed that several stellar population components contribute to the heating of the dust, and conversion between infrared flux and SFR is likely to change in different types of galaxies (e.g., Cox et al. 1986, Gallagher and Hunter 1986, Persson and Helou 1986, Walterbos and Schwering 1986). The enormous potential of the IRAS data base for this problem has stimulated a vigorous effort to understand and calibrate an IR-based SFR scale, however, and significant progress is expected in the near future.

Resolved Stars: This is clearly the most direct method for deriving SFRs in galaxies, and stellar statistics have been applied to study the star formation histories of a few nearby galaxies, (e.g., Lequeux 1979). Aside from a few dwarfs in the Local Group, however, the stellar data are incomplete, especially at the very high-mass end (e.g., Massey 1985), preventing detailed study of integrated star formation properties. On-going surveys of the Local Group members by Massey, Garmany, DeGioia-Eastwood, Conti, Walborn, and others promise to rectify this situation in the near future.

All of these methods are potentially useful for studying star formation histories, but at present only the Lyman photon-counting measurements have been applied for this purpose, for the Galaxy (Güsten and Mezger 1983), and for a large sample of spiral and irregular galaxies (Kennicutt 1983, Gallagher et al. 1984). These three studies have yielded very similar results, and I shall briefly review the Hα work as an example.

Figure 6 shows another two-color diagram for spiral galaxies; in this case the ordinate is not U-B but rather the equivalent width of the H-alpha emission line in the integrated spectrum, the total Balmer luminosity normalized to the red continuum luminosity of each galaxy. This is effectively a color index for the galaxy between 900Å and 6600Å, and as such is sensitive to both the ratio of present to past star formation as well as to the slope of the IMF above ~1 $M_\odot$. Superimposed are the same "standard" galaxy evolution models which were shown earlier in Figure 2 (the computations were extended to the Lyman continuum in order to synthesize the Balmer equivalent widths). The H-alpha emission is quite sensitive to the IMF, and this relationship can be used to constrain the IMF slope in the galaxies, and to estimate total SFRs from the Balmer fluxes. For details see Kennicutt (1983).

Figure 6 indicates that the same birthrate model which can account for the UBV colors of disk galaxies also provides an excellent fit to the observed Balmer fluxes. This point is demonstrated directly in Figure 7, which compares the birthrate histories as inferred from the H-alpha derived SFRs with those derived from color models (e.g., Figure 3). The general agreement is excellent. Figure 7 also provides

Figure 6.   Observed H-alpha emission line equivalent widths versus color for field spirals, from Kennicutt (1983).   The 3 shaded areas are the same evolutionary models as shown in Figure 2.   The shading denotes uncertainty due to extinction of the Balmer emission and reddening of the colors.

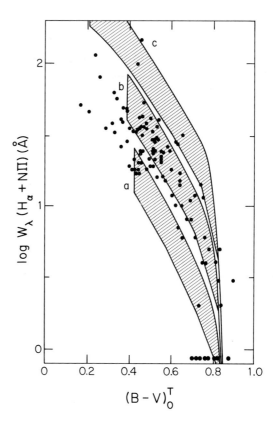

Figure 7.   Comparison of birthrate history parameter b as derived from broadband color models and from $H\alpha$ determined star formation rates.   The computations for early-type spirals are less reliable, due to bulge contamination and extinction. Arrows denote $H\alpha$ upper limits.

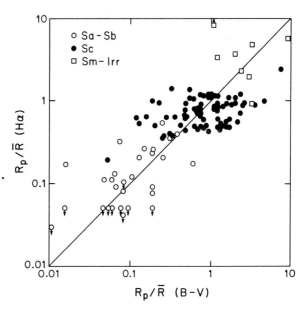

a useful external estimate of the accuracy of the star formation
history derived for individual galaxies using either method. The
considerable scatter confirms our earlier suspicions about the likely
uncertainties in the color models, but the H-alpha observations do
appear to confirm the general conclusions drawn earlier. For a
thorough comparison of the models and discussion, see Gallagher et al.
(1984).

Summary

In this section I have discussed several different observational
approaches to constraining the star formation histories of galaxies,
and all of them are consistent with a common basic birthrate picture.
Specifically:

1) All integrated photometric properties of galaxies from the Lyman
continuum to the red can be reproduced with a simple one-parameter
sequence of model galaxies with the same age, metallicity, and IMF, and
with the birthrate history as the primary variable. This does not
necessarily imply that variations in the other parameters are unimportant; metallicity variations, for example, are probably very important
in the near-infrared (Bothun et al. 1984).

2) The birthrate histories implied by this simple model are
relatively constant in spiral disks. The SFR in a typical Sc disk has
been roughly constant since its formation; in most Sa-Sb disk the
current SFR is a few tenths of its average value in the past. In many
galaxies the global SFR may actually increase with time. Studies of
the stellar population in the solar neighborhood are consistent with
this result.

3) Fewer data are available on the relative variation of the
birthrate history within the disks of individual galaxies, but
preliminary results suggest that the histories are relatively uniform
spatially, as well as temporally. Studies of the radial distributions
of resolved blue stars (Freedman 1984) and HII regions (Hodge and
Kennicutt 1983, Kennicutt et al., in preparation) in nearby spirals
indicate that the distribution of the young starlight closely tracks
that of the integrated light from the old disk in most (but not all!)
late-type spirals. An example of this behavior is illustrated in
Figure 8. The relative uniformity of disk colors in most spiral
galaxies (e.g., Schweizer 1976, Weavers 1984) is consistent with this
interpretation.

4) The most important caveat. Most tracers of the star formation
histories in galaxies do not measure the integrated birthrate history
directly, but rather the ratio of the present SFR of massive stars to
the past SFR of low-mass stars. This is a fundamental limitation
imposed by nature (there are no old massive stars!), and unfortunately
it means that SFR data alone cannot distinguish unambiguously between a
constant IMF, evolving birthrate model, and an evolving IMF model such

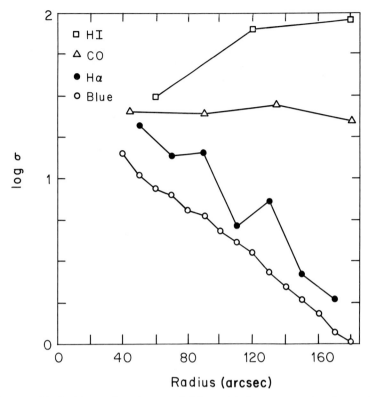

Figure 8.   Radial distributions of blue light, Hα emission, HI column density, and CO molecular line intensity in the Sb galaxy NGC 2841. Note the close correspondence between the Hα and broadband continuum profiles, and the poor correlation between Hα emission and either of the gaseous components.

as Larson's bimodal picture.   It also limits the accuracy with which the birthrate history can be derived for individual galaxies to the factor-of-two level at best; but the overall consistency of the different methods suggests that the general picture outlined here is correct.

COMPARISON WITH SIMPLE EVOLUTION MODELS

     These relatively constant birthrates are not understood theoretically.   In order to illustrate the point, consider a simple closed-box model for the evolution of a "perfect" galaxy, where the SFR varies as a simple power of the gas density, the well-known Schmidt (1959) law:

$$R(t) = A\rho^n \tag{3}$$

The explicit solution for the evolution of the gas density ρ and the star formation rate R will depend on the power-law index   n   and on the boundary conditions. For our perfect galaxy we adopt the simplest possible boundary conditions, a closed system (i.e., no gaseou0 infall or outflow), and a time-independent geometry for the gas and stars. The latter assumption allows us to parametrize the time dependence of the SFR in terms of the global properties of the gas, rather than the local density.

Although this heuristic model for the evolution of a galactic disk is almost certainly unrealistic in detail, it is probably the most widely applied model for the chemical evolution of disks, and the properties of the solution of this model have been discussed in several review articles (e.g., Audouze and Tinsley 1976, Güsten and Mezger 1983). Under the boundary conditions applied here, the gas density will decline either exponentially with time (n=1), or as a power law in time (n>1), with the SFR declining accordingly. In any case, the SFR at any time, normalized to the initial SFR, depends only on the fractional gas content of the disk:

$$\frac{R(t)}{R_o} = (\frac{\rho(t)}{\rho_o})^n \tag{4}$$

How well do real galaxies conform to this simple prediction? Figure 9 shows the relationship between the normalized SFR and fractional gas mass for the spiral and irregular galaxies in the Kennicutt (1983) sample. The SFR ratio has been extrapolated from the b values shown in Figure 7, using an exponential time dependence; for most galaxies the SFR is relatively constant, so the extrapolation is not very large. The fractional gas masses have been estimated from the measured HI masses, multiplied by a factor three to approximately correct for molecular gas and helium, and normalized to the stellar mass of disk within the RC2 radius (de Vaucouleurs et al. 1976). (The same disk mass is used in the SFR normalization; adopting a different radius will move points along the arrow shown in Figure 9.) Under these assumptions I have probably overestimated the fractional gas contents of most of the galaxies; this was done intentionally to bias the comparison in favor of the simple Schmidt model. Also note that in many galaxies much of the gas lies outside of the active star forming disk, so again our comparison is probably biased in favor of the model.

As illustrated in Figure 9, the star formation histories of most galaxies are much flatter than would be expected from a simple SFR vs. gas content scaling law. The current SFRs are several times higher than predicted from a linear Schmidt law, and 1-2 orders of magnitude higher than expected from a quadratic law. The Galactic disk is no exception; I have indicated with a box the range of values derived by Twarog (1980) and Güsten and Mezger (1983) for the solar neighborhood and the Galactic disk.

A time-averaged version of this birthrate problem is the gas consumption "problem" for spirals. As pointed out by Larson et al. (1980) and by Kennicutt (1983), most disk galaxies will exhaust their

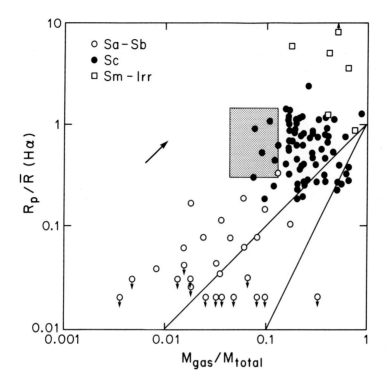

Figure 9.   Present SFR normalized to the initial SFR, as estimated from
Hα emission and an exponential SFR history, plotted as a function of
fractional gas mass, for a sample of spiral and irregular galaxies.
The box indicates the range of parameters for the Galactic disk, from
various estimates.   The lines indicate the expected relations for
Schmidt star formation laws with power-law exponents of 1 and 2.
Adopting a different disk radius moves points along the direction of
the arrow.

gas supplies over the next 5 Gyr if they continue to form stars at their
present rates.   Many of the proposed modifications to the simple model
discussed in the next section would increase the future time scales for
star formation in galaxies as well.

IMPLICATIONS

     These apparent discrepancies between the observed and predicted
star formation histories of galaxies are nothing new.   Schmidt (1962)
encountered the birthrate problem in the solar neighborhood, and
numerous investigators since then have discussed the problem and offered
possible solutions.

The discrepancies could arise from three sources, errors in the observations, errors in the boundary conditions applied to the simple star formation model, or fundamental errors in the model itself. Suspicions have been raised in the literature toward all three sources, and in this section I briefly review the possible alternatives.

Observations?

I attempted to present a critical assessment of the relevant observations in section two. In summary, although it is possible that the "observed" birthrate histories may possess systematic errors as large as factors of 2-3, it is extremely unlikely that observational errors alone could account for the much larger discrepancies between the inferred star formation histories and the predictions of the simple gas content model.

As discussed earlier, however, the interpretation of the SFR data in terms of star formation histories requires certain assumptions about the evolution of the stellar population, in particular the assumption that the IMF has not evolved appreciably over the lifetime of the disk. Existing observations only constrain the birthrate history of low-mass stars, the present-day IMF, and to some extent the ratio of current massive star formation to past low-mass star formation. Within these loose observational constraints it is possible to devise models with bimodal and time-dependent IMFs which can fit the current observations (including chemical enrichment constraints) and avoid the birthrate and gas consumption problems. Jensen et al. (1981), Güsten and Mezger (1983), and Larson (1986) have proposed bimodal IMF models which specifically address the constraints imposed by the star formation histories. The reader is referred to Larson's review in this volume for a thorough discussion of this models.

As pointed out by Larson (1986), such a bimodal IMF model can incorporate a power-law dependence of the SFR on gas density (a different dependence for each of the two IMF components) without running afoul of the observational constraints. The simple star formation law in equation (3) becomes much more complicated, however:

$$R(t) = A(t)\rho^{n_1} + B(t)\rho^{n_2}$$

where A and B denote the initial normalizations of the two IMF components, and n_1 and n_2 denote the different dependences on density. Although the bimodal IMF models have been included under the heading of questionable interpretation of the observations, they obviously entail a major retooling of the basic galactic evolution model as well. Unfortunately, even in this relatively simple case, there are more additional free parameters than can be constrained by the available observations.

The only other critical observational parameter is the fractional gas mass of the galaxies. If the contributions of molecular gas to the total gas content (assumed to be roughly half the total on average) were grossly underestimated, it might be possible to reconcile the observations with at least a shallow SFR-density power law (n=1). Such large underestimates are unlikely, however (Verter 1986).

Boundary Conditions?

The two most important boundary conditions in the model were the assumptions of a closed system and of a fixed geometry for the gas. Both assumptions are questionable.

Infall of gas on to the disk was originally invoked in an attempt to account for the metallicity evolution of the galactic disks (Larson 1972), but it clearly acts to flatten the star formation history as well. Infall models which have been directed specifically at the birth-rate problem include those by Ostriker and Thuan (1972), Larson et al. (1980), Güsten and Mezger (1983), and Lacey and Fall (1985). Star formation in early-type spirals or in the Galaxy can be sustained with relatively modest levels of infall, but in late-type spirals and irregulars, where the stellar birthrate is constant or even increasing with time, the infall rates would have to be prodigious. Such large flows are not observed, and in any case are difficult to reconcile with other observed properties of the galaxies (Kennicutt 1983). More modest infall levels could be very effective, however, when combined with one or more of the other mechanisms discussed here (e.g., Güsten and Mezger 1983, Lacey and Fall 1985, Shore et al. 1986).

Evolution in either the vertical scale height or the radial distribution of the gas in a disk may also moderate the birthrates in disks. Twarog (1980) finds evidence for a steady decrease in the scale height of the disk over its history, and this could slow the drop in the SFR with time, especially if the Schmidt power law index n were large. Lacey and Fall (1985) have studied disk evolution models with modest radial gas flows, and have found that such models can fit both the chemical and birthrate constraints in the Galaxy, especially if the radial flows are combined with infall from the halo.

Star Formation Model?

Most of the work described so far has been directed at relaxing the boundary conditions of the basic model, while maintaining the Schmidt-type star formation law. Several recent studies, however, have raised suspicions about the general validity of the SFR-density power law as a reliable description of the global birthrate in galaxies.

The observational evidence for the Schmidt law has been reviewed by Madore (1977) and Freedman (1984). It consists mainly of observed correlations between the column densities of HI gas and the local surface densities of resolved stars or HII regions, following procedures first outlined by Schmidt (1959) and Sanduleak (1969). The values of n derived in such studies have ranged from about 0.5 to 3.5, with an

average of 1-2. Several investigators have also examined the
correlation between the SFR and the intensity of the CO molecular line;
in this case the exponent is usually close to unity (e.g., Talbot 1980,
Young and Scoville 1982).

While these studies have confirmed the existence of a positive
correlation between the local SFR and gas density in some regions, the
physical interpretation of these measurements has been called into
question by several workers, including Madore (1977), Buibert et al.
(1978), and Freedman (1984). Experiments with high resolution data
(Freedman 1984) and with numerical simulations (Madore 1977, Freedman
1984) have demonstrated that the slope of the observed SFR-density
correlation is extremely sensitive to the spatial resolution employed,
and in most cases is much steeper than the true physical relation.
Allen et al. (1986), and Shaya and Federman (1986) have gone so far as
to suggest that the physical cause-and-effect relationship between
average HI density and the SFR is the reverse of what is commonly
assumed, if the primary source of HI is dissociation of molecular gas
by young stars.

For the galactic evolution problem we are more interested in the
dependence of the large-scale star formation activity on the available
gas supply, rather than on the local relationship, which probably tells
us more about the efficiency of star formation once it is under way.
The validity of the Schmidt law in this regime is even more suspect,
however. Figure 8 shows the radial variations in H-alpha emission, HI
column density, and CO intensity in NGC 2841, a nearby Sb galaxy. Over
most of the optical disk of the galaxy the SFR appears to be anticorrelated with both interstellar gas components! I have chosen this
galaxy as an extreme case, but in a more general study of 20 spirals
(Kennicutt et al. 1986), we find that there is virtually no correlation
between the large-scale HI density and the massive SFR, aside from a
rough correlation which shows up when the entire galaxy is averaged
together (see also Guiderdoni and Rocca-Volmerange 1985). In some
galaxies there appears to be a tighter correlation of the SFR with CO
intensity (e.g., Young and Scoville 1982), but in others the correlation
is altogether absent. While these results do not necessarily invalidate
the Schmidt law as a useful parametrization of the star formation on
small scales, they do raise serious concerns about its use in galaxy
evolution models.

What are the alternatives? As pointed out by Lynden-Bell (1977),
the Schmidt law was never intended to be a precise physical description
of the star formation process, and a rigorous physical model would
probably be so complicated as to be rendered useless for galaxy models.
Over the past few years, however, several workers have offered alternative parametrizations for the regulation of the stellar birthrate which
attempt to avoid some of the major physical shortcomings of the Schmidt
law, while retaining most of the simplicity which makes it so useful
for galaxy modelling. A comprehensive summary of this work is beyond
the scope of this review (see Elmegreen 1985 instead), but a couple of
examples will serve to illustrate the general direction of this research.

At very low gas densities the Schmidt law must break down, because the gas will no longer be gravitationally unstable. For spirals the critical density is set by rotational shear or by the galactic tidal field. As pointed out by Quirk (1972) and Elmegreen (1979), these threshold densities are comparable to the actual interstellar gas densities over substantial regions of many galaxies, and in those regions star formation will be inhibited or at least segregated to regions of global gas compression, such as spiral arms. Such threshold densities for star formation are actually observed in irregular galaxies (Davies et al. 1976, Hunter and Gallagher 1986, Skillman 1986) and in spirals (van der Hulst et al. 1986, Kennicutt et al. 1986). In the latter the levels are roughly consistent with those expected from the simple shear picture.

At relatively high SFRs we might also expect the Schmidt law to break down, as the young stars disrupt the interstellar medium in which they formed. Several self-regulating or feedback models for star formation in the Galaxy and in external galaxies have been published recently (e.g., Cox 1983, Franco and Cox 1983, Franco and Shore 1984, Ikeuchi et al. 1984, Dopita 1985, Scalo and Struck-Marcell 1986). In most the star formation is limited by equilibrium between energy input to the ISM from supernovae and stellar winds, balanced by the confining pressure of the galactic ISM. While most of these models are too complicated at present to compare directly with the observations, they do reproduce many of the qualitative features of star formation in galaxies, as well as the relatively flat birthrates which are observed.

Summary and Concluding Remarks

As must be apparent by now, there is no shortage of possible solutions to understanding the observed star formation histories in galaxies. The problem at present is not the lack of an answer, rather the task of sorting through the available answers and isolating the most important effects. The recent progress in the observations of star formation rates in galaxies should aid greatly in this respect. Although our present state of knowledge may appear to be ambiguous and confused (it is), the field is maturing rapidly. The eventual product of this effort should be a more mature and physical understanding of the evolution of galaxies.

REFERENCES

Aaronson, M. 1978, Ap.J. (Letters), 221, L103.
Allen, R.J., Atherton, P.D., and Tilanus, R.P.J. 1986, Nature, 319, 296.
Audouze, J., and Tinsley, B.M. 1976, Ann. Rev. Astr. Ap., 14, 43.
Butcher, H. 1977, Ap.J., 216, 372.
Cox, D.P. 1983, Ap.J. (Letters), 265, L61.
Cox, P.N., Kügel, E., and Mezger, P.G. 1986, Astr. Ap., 155, 380.
Davies, R.D., Elliot, K.H., and Meaburn, J. 1976, Mem.R.A.S., 81, 89.

de Vaucouleurs, G., de Vaucouleurs, A., and Corwin, H.G. 1976, Second Reference Catalog of Bright Galaxies, Austin: University of Texas Press.
Donas, J., and Deharveng, J.M. 1984, Astr. Ap., 140, 325.
Dopita, M.A. 1985, Ap.J. (Letters), 295, L5.
Elmegreen, B.G. 1979, Ap.J., 231, 372.
Franco, J., and Cox, D.P. 1983, Ap.J., 273, 243.
Franco, J., and Shore, S.N. 1984, Ap.J., 285, 813.
Freedman, W.L. 1984, Ph.D. Thesis, University of Zoronto.
Gallagher, J.S., and Hunter, D.A. 1986, in Star Formation in Galaxies, ed. N. Scoville and G. Neugebauer, in press.
Gallagher, J.S., Hunter, D.A., and Tutukov, A.V. 1984, Ap.J., 284, 544.
Gehrz, R.D., Sramek, R.A., and Weedman, D.W. 1983, Ap.J., 267, 551.
Guibert, J., Lequeux, J., and Viallefond, F. 1978, Astr. Ap., 68, 1.
Güsten, R., and Mezger, P.G. 1983, Vistas Astr., 26, 159.
Hardy, E., Buonanno, R., Corsi, C.E., Janes, K.A., and Schommer, R.A. 1984, Ap.J., 278, 592.
Hodge, P.W., and Kennicutt, R.C. 1983, Ap.J., 267, 563.
Houk, N. 1982, Michigan Spectral Catalog, Ann Arbor: University of Michigan.
Hunter, D.A., and Gallagher, J.S. 1986, P.A.S.P., in press.
Hunter, D.A., Gillett, F.C., Gallagher, J.S., Rice, W.L., and Low, F.J. 1986, Ap.J., 303, 171.
Ikeuchi, S., Habe, A., and Tanaka, K. 1984, M.N.R.A.S., 207, 909.
Jensen, E.B., Talbot, R.J., and Dufour, R.J. 1981, Ap.J., 243, 716.
Kennicutt, R.C. 1983, Ap.J., 272, 54.
Kennicutt, R.C., Edgar, B.K., and Hodge, P.W. 1986, in preparation.
Kennicutt, R.C., and Kent, S.M. 1983, A.J., 88, 1094.
Klein, U., Wielebinski, R., and Beck, R. 1984, Astr. Ap., 135, 213.
Kron, R.G. 1982, Vistas Astr., 26, 37.
Lacey, C.G., and Fall, S.M. 1985, Ap.J., 290, 154.
Larson, R.B. 1972, Nature, 236, 21.
Larson, R.B. 1986, M.N.R.A.S., 218, 409.
Larson, R.B., and Tinsley, B.M. 1978, Ap.J., 219, 46.
Larson, R.B., Tinsley, B.M., and Caldwell, C.N. 1980, Ap.J., 237, 692.
Lequeux, J. 1979, Astr. Ap., 71, 1.
Lequeux, J. 1985, in Spectral Evolution of Galaxies, ed. C. Chiosi and A. Renzini, Dordrecht: Reidel.
Lynden-Bell, D. 1977, in IAU Symposium No. 75, Star Formation, ed. T. de Jong and A. Maeder, Dordrecht: Reidel, p. 291.
Madore, B.F. 1977, M.N.R.A.S., 177, 1.
Massey, P. 1985, P.A.S.P., 97, 5.
Miller, G., and Scalo, J. 1979, Ap.J. Suppl., 41, 513.
Ostriker, J.P., and Thuan, T.X. 1975, Ap.J., 202, 353.
Persson, C.J., and Helou, G. 1986, Ap.J., submitted.
Quirk, W.J. 1972, Ap.J. (Letters), 176, L9.
Rocca-Volmerange, B., Lequeux, J., and Maucherat-Jobert, M. 1981, Astr. Ap., 104, 177.
Sanduleak, N. 1969, A.J., 74, 47.
Scalo, J.M. 1986, Fund. Cosmic Phys., in press.
Scalo, J.M., and Struck-Marcell, C. 1986, Ap.J., in press.

Schmidt, M. 1959, Ap.J., 129, 243.
Schmidt, M. 1963, Ap.J., 137, 758.
Schweizer, F. 1976, Ap.J. Suppl., 31, 313.
Searle, L., Sargent, W.L.W., and Bagnuolo, W.G. 1973, Ap.J., 179, 427.
Shaya, E.J., and Federman, S.R. 1986, Bull.A.A.S., 18, 707.
Shore, S.N., Ferrini, F., and Palla, F. 1986, Ap.J., in press.
Skillman, E.D. 1986, Bull.A.A.S., 18, 691.
Smith, L.F., Biermann, P., and Mezger, P.G. 1978, Astr. Ap., 66, 65.
Stryker, L.L., and Butcher, H.R. 1981, in IAU Colloquium No. 68,
    Astrophysical Parameters for Globular Clusters, ed. A.G.D. Philip
    and D.S. Hayes, Schenectady: L. Davis Press, p. 225.
Talbot, R.J. 1980, Ap.J., 235, 821.
Thronson, H.A., and Telesco, C.M. 1986, Ap.J., in press.
Tinsley, B.M. 1968, Ap.J., 151, 547.
Tinsley, B.M. 1972, Astr. Ap., 20, 383.
Tinsley, B.M. 1977, Ap.J., 216, 548.
Tinsley, B.M. 1980, Fund. Cosmic Phys., 5, 287.
Twarog, B.A. 1980, Ap.J., 242, 242.
van der Hulst, J.M., Skillman, E., Kennicutt, R., and Bothun, G. 1986,
    Astr. Ap., submitted.
Vangioni-Flam, E., Lequeux, J., Maucherat-Joubert, M., and Rocca-
    Volmerange, B. 1980, Astr. Ap., 90, 73.
Verter, F. 1986, preprint.
Walterbos, R.A.M. 1986, Ph.D. Thesis, Leiden University.
Weavers, B.M.H.R. 1984, Ph.D. Thesis, Groningen University.
Young, J.S., and Scoville, N.Z. 1982, Ap.J., 258, 467.

DISCUSSION

BURSTEIN: Most of the interpretations of stellar properties and motions
near the sun assume that they are typical of a) our galaxy and
b) galaxies in general. Yet, we know from interstellar medium studies
that we live in an evacuated bubble; and from the existence of Gould's
belt, that star formation can exist in rather peculiar configurations
with respect to us. In this context, I have to wonder: when comparing
the properties of young ($\leq 50 \times 10^6$ yrs) stars to old stars, what is
ubiguitous (i.e., general to the galaxy), and what is unique to us?

KENNICUTT: Twarog studied the history of star formation over billion-
year time scales in the solar neighborhood, so there the local effects
should average out. Güsten and Mezger have measured the massive star
formation over much larger scales in the Galaxy. You are certainly
correct in that a study of the star formation history of, for example,
the molecular ring would yield very different results from one in the
solar neighborhood, and one must average over a representative spatial
scale.

ROCCA-VOLMERANGE: About the star formation regulation process, I have
to mention a recent paper submitted to A. & A. by B. Guiderdoni in
which an observational threshold of gas surface density is evidenced in

a sample of Virgo Cluster anemic galaxies. The value of such a
threshold is quite in agreement with the star formation models of
Elmegreen 1981 and Dopita 1985. Moreover, estimates of the $R_{present}/R_{past}$ ratio in spiral galaxies are strongly affected by reddening. Have
you an idea about the respective uncertainties of dereddening if the Hα
or far-UV estimators are used?

KENNICUTT: The typical extinction at Hα is about a magnitude, with a
large dispersion about that mean. Extinction of the Hα emission line
itself will be much lower than the UV emission of course, and it can be
determined directly from the radio or optical spectra of the HII
regions. It is possible, however, that significant extinction occurs
within the star forming regions, and absorbs ionizing photons before
they are absorbed by the gas. In that case the Balmer and UV
extinctions will be comparable.

ZINNECKER: Can you say anything about the star formation history of SO
galaxies?

KENNICUTT: Most of the best constraints on star formation in SO
galaxies are based on spectral synthesis observations, and these will
be reviewed in O'Connell's talk. I believe the most recent studies are
by Gregg and Caldwell. Both find evidence for low levels of "recent"
star formation in some SO disks. There is also evidence for nuclear
star formation in some systems (see, e.g., the poster here by
Rocca-Volmerange et al.).

WHITMORE: You indicated that the star formation rate in interacting
galaxies was much higher than in normal galaxies. Since we can make a
reasonable estimate of the age of these systems from dynamical arguments, it would be easy to compare the gas used in the recent starburst
to the total amount of gas. Is the "gas consumption" problem for these
systems better or worse than for normal galaxies?

KENNICUTT: The star formation rates in the interacting galaxies show
an enormous dispersion. The most extreme cases, with star formation
rates 10-100x normal, would presumably exhaust their gas supplies in
times of order 10^8 years; hence the bursts must be short. Most of the
interacting galaxies have emission levels which are within a factor 2-3
of normal galaxies, however, and the gas consumed in the interaction
will be negligible. Also note that if the IMF in the starbursting
systems is heavily biased to massive stars, as is suspected by several
workers, the gas consumption rates will be lower than you would expect
from looking at the emission luminosities.

BOND: Nancy Houk's H-R diagram shows two distinct peaks on the main
sequence. Is there an explanation for this structure?

ROMAN: It must be remembered that spectral type boxes are non-uniform
in size (e.g., mass intervals). Thus the shape of Nancy Houk's diagram
is the result of the interplay of the distance to which you see stars

and the size of the spectral type box, as well as of the actual stellar mass distribution.

KENNICUTT: Certainly correct. Faint stars are obviously undersampled in Figure 4, but since the giants and dwarfs which dominate the light of galaxies are of comparable luminosity, it turns out not to affect the conclusions drawn here.

YIELD AND ABUNDANCE CONSTRAINTS ON
GALACTIC CHEMICAL EVOLUTION

James W. Truran
Department of Astronomy, University of Illinois

Friedrich-Karl Thielemann*
Department of Astronomy, University of Illinois

Abstract.   A brief review is presented of the nucleosynthesis   mechanisms   which   have   been   found   to   be
responsible   for   the   formation   of   most   of   the   heavy
elements   present   in   galactic   matter.   The   pertinent
astrophysical   environments   are   identified   as   being
associated   with   the   evolution   of   intermediate   mass   and
massive stars and with supernovae of types I and II, and
specific predictions of the yields for these astrophysical sites are summarized.   We survey significant trends
in galactic abundances, including abundance patterns in
extremely metal-deficient halo field stars and in globular cluster stars and the metallicity distribution for
disk stars.   We then seek to identify constraints that
these combined observational and theoretical considerations   impose   upon   theories   of   the   formation,   early
history, and evolution of our galaxy.

INTRODUCTION

        Studies of galactic chemical evolution seek to explain
the distributions of abundances of the elements observed in the stars
and interstellar matter in galaxies.   Such studies require that one
provide some measure of the rate of star formation (SRF) as a function of time, of the stellar initial mass function (IMF) as a function of time, of the fraction of gas processed through stars, and of
the evolutionary behaviors of stars of different masses as a function
of their initial compositions and their implications for nucleosynthesis.   Such factors are expected strongly to influence the yields
of stellar generations and therefore the abundance history of the
matter of which a galaxy is composed.   In turn, scrutiny of the
abundance patterns observed in stars of various ages (metallicities)
can provide clues to and constraints upon the history of stellar
activity and nucleosynthesis.   An excellent review of problems of
galactic chemical evolution has been provided by Tinsley (1980).

*On leave from the Max-Planck-Institut für Physik und Astrophysik,
Garching bei Munchen, FRG

Questions concerning the histories of star formation activity in galaxies and possible variations in the form of the initial mass function are addressed elsewhere in these proceedings. My aim in this paper is to focus upon two other aspects of this problem - nucleosynthesis yields and abundance trends in the stars and gas of our galaxy - and to demonstrate how the combination of these ingredients allows one to identify constraints on galactic evolution. A review of basic nucleosynthesis mechanisms, the astrophysical sites in which they are believed to operate, and predicted nucleosynthesis yields as a function of stellar mass is presented in the next section. The accepted view that the synthesis of the bulk of the elements heavier than helium is a consequence of nuclear processes occurring in stars (primarily more massive stars of shorter lifetimes) over the history of our galaxy implies that a monotonic increase in the metal content (Z) of the gas should characterize early galactic history. One might expect there to be differences in relative abundances of nuclei formed by different nucleosynthesis processes, in stellar or supernova environments provided by stars of different masses (lifetimes), superimposed upon the overall increasing trend and, indeed, such abundance variations do exist (Truran 1983, 1984b). Observations of significant elemental abundance patterns and trends in the most metal-deficient field halo stars in our galaxy, in stars in globular clusters, and in disk population stars are summarized and discussed. The possible implications of these observations and the constraints they serve to impose upon, particularly, the earliest stages of evolution of our galaxy are then identified and examined.

NUCLEOSYNTHESIS PROCESSES, SITES, AND YIELDS

Critical input to studies of galactic chemical evolution is provided by calculations of stellar and supernova nucleosynthesis. Reviews of nucleosynthesis theory have been provided by Arnett (1973), Truran (1973a), Trimble (1975), and Truran (1984a). For the purposes of our present discussion, we will identify groups of elements which are believed to have common origins in astrophysical environments, review the distinguishing characteristics of the nuclear processes which are assumed to be responsible for their formation, and seek to identify the astrophysical sites - in stars and in supernovae of Types I and II - in which these nucleosynthesis mechanisms are believed to operate.

Hydrogen, helium, lithium, beryllium and boron

The isotopes of the light elements H, He, Li, Be, and B are understood to have been formed in two quite different astronomical environments. The observed abundances of ^6Li, ^9Be, ^{10}B, and ^{11}B are entirely consistent with the view that they have been formed by the interactions of cosmic rays with the constituents of the interstellar medium over the course of galactic evolution (Reeves et al 1973; Reeves 1974). The nuclei H, D, ^3He, ^4He, and ^7Li constitute the major products emerging from the cosmological big bang (Boesgaard and Steigman 1985). If one assumes the validity of the standard big bang model, the observed abundances of these isotopes in galactic

matter can in principal be used to impose constraints upon both cos-
mology and particle physics (Yang et al 1984). ^{4}He is considered to
be largely cosmological in origin, with only a small level of production in the galaxy. ^{2}D is readily destroyed in stellar interiors and
there is no recognized galactic source for its production, so its
abundance in the interstellar medium today is presumed to represent a
lower limit on its primordial value.   Numerical models of galactic
evolution (Truran and Cameron 1971; Tinsley 1980) suggest a reduction
in deuterium of a factor of ~ 2-3.   The situations for ^{3}He and ^{7}Li
are complicated by the fact that these nuclei can be both formed and
destroyed in different stellar environments.   The low lithium abundance ^{7}Li/H = 1.12×10^{-10} recently determined for unevolved halo stars
and old disk stars in our galaxy by Spite and Spite (1982) may provide a reasonable appraisal of the primordial ^{7}Li concentration, but
there remain serious questions regarding this interpretation
(Michaud, Fontaine, and Beaudet 1984).   In light of the known uncertainties, the general consistency of the observed abundances with
predicted cosmological results is at once suggestive and encouraging.

The elements carbon, nitrogen, and oxygen
The seven stable isotopes of carbon, nitrogen, and oxygen
are known to have a quite complex nucleosynthesis history (Truran
1977; Audouze 1985; Diaz and Tosi 1986).   We note in particular that
the abundances of the rarer isotopes ^{13}C, ^{15}N, ^{17}O, and ^{18}O may
include interesting contributions from red giant stars (Renzini and
Voli 1981), novae (Truran 1985), and supernovae.   For the purposes of
this paper, we will be concerned only with the modes of production of
the dominant isotopes ^{12}C, ^{14}N, and ^{16}O.

^{14}N is formed as a consequence of CNO-cycle hydrogen
burning.   The astrophysical site is generally assumed to be the
hydrogen burning shells of red giant stars where the ^{14}N thus formed
can subsequently be transported to the surface by convection and
enrich the interstellar medium as a consequence of mass loss or
planetary nebula ejection.   In general, all primordial CNO nuclei in
the shell will be converted to ^{14}N due to these burning sequences:
the ^{14}N thus formed therefore represents a "secondary" nucleosynthesis product (it demands the input of metals from an earlier stellar
generation).   Mixing of the products of shell helium burning into the
overlying hydrogen envelope during the course of the red giant evolution of intermediate mass stars allows for the production of ^{14}N as a
primary nucleosynthesis product (Truran and Cameron 1971; Iben and
Truran 1978; Renzini and Voli 1981).   Unfortunately, the considerable
uncertainties associated with predictions of nitrogen production
preclude the use of observations of ^{14}N abundances as effective
constraints on galactic chemical evolution.

^{12}C and ^{16}O constitute the primary products of helium
burning in stars.   The ^{12}C/^{16}O ratio at the end of helium burning is
a function of the effective rate of the reaction ^{12}C$(\alpha,\gamma)^{16}$O during
helium burning, which itself is dependent upon both the intrinsic
rate properties and the prevailing burning temperature.   Some effects
of variations in the rate of the ^{12}C$(\alpha,\gamma)^{16}$O reaction on the chemical
evolution of the solar neighborhood have been reviewed by Matteucci

(1986). For the purposes of this paper, we will note only the following quite general trends. The $^{16}O/^{12}C$ ratio characterizing matter processed through massive stars ($M \geq 10$ $M_\odot$) is high relative to solar system matter (Arnett 1978; Weaver, Zimmerman, and Woosley 1978; Woosley and Weaver 1982; Arnett and Thielemann 1985; Woosley and Weaver 1986). This is a consequence both of the (current) rate of the $^{12}C(\alpha,\gamma)^{16}O$ reaction and of the increasing effectiveness of this reaction relative to the 3 $^4He \rightarrow {}^{12}C$ reaction as the ^4He abundance decreases during the final stages of helium burning. Intermediate mass stars ($2 \lesssim M \lesssim 8$-10 $M_\odot$) on the asymptotic giant branch may alternatively provide the source of ^{12}C necessary to meet galactic requirements (Truran 1973b; Iben and Truran 1978; Renzini and Voli 1981). Here, ratios $^{12}C/^{16}O > 1$ typically arise due to the outward mixing of products of incomplete helium burning from the thermally pulsing shells of these luminous red giants.

The elements neon to nickel

Most of the nuclei present in nature in the mass range $20 \lesssim A \lesssim 60$ are formed collectively as a consequence of the processing of the cores of massive stars ($M \geq 10$ $M_\odot$) during the late stages of presupernova evolution and/or of explosive nucleosynthesis accompanying mass ejection in the ensuing Type II supernova events (Arnett 1978; Weaver, Zimmerman, and Woosley 1978; Woosley and Weaver 1982; Thielemann and Arnett 1985; Woosley and Weaver 1986). Successive exoergic stages of burning of hydrogen, helium, carbon, oxygen, and silicon fuels define the presupernova evolution of these massive stars. When the ashes of these burning epochs are subsequently subjected to high temperatures and densities accompanying their ejection in supernova events, further thermonuclear processing occurs. The elemental and isotopic abundance patterns resulting from these burning phases strongly resemble those of solar system matter.

A somewhat more detailed review of massive star nucleosynthesis and its implications for the yields of stellar populations is presented by Olive, Thielemann and Truran (1986). The following general trends are of particular importance to our subsequent discussions and are relatively independent of differences in detail between the yield predictions of different researchers. Models for the more massive stars ($M > 10$ $M_\odot$) generally predict relatively high ratios O/Fe compared to solar system matter; the O/C ratio is correspondingly high. These trends tend to carry over to the alpha-particle nuclei of intermediate mass: ^{20}Ne, ^{24}Mg, ^{28}Si, ^{32}S, ^{36}Ar, and ^{40}Ca are all somewhat overproduced relative to Fe. Although the mass of iron ejected in such a supernova event is highly uncertain, due to uncertainties associated with the position of the mass cut and with the remnant masses, the tendency for iron to be underproduced seems clear. It would thus appear likely that the abundances both of oxygen and of the elements in the range Ne - Ti in galactic matter are dominated by contributions from massive stars and associated Type II supernovae. The details will of course be dependent upon the temperature and density structure of the presupernova star, so that the ejecta of supernovae of different masses may be expected to differ somewhat in the relative abundances of elements produced, but the integrated contributions from massive stars are probably reasonably well known. Furthermore, calculations indicate that the abundances both of odd-Z nuclei and of the neutron-rich isotopes of even-Z nuclei may be significantly depleted in matter ejected by extremely metal-deficient stars (Truran and Arnett 1971); some odd Z - even Z elemental abundance variations may thus be expected for metal-poor stars.

Numerical studies currently suggest that the contributions from Type I supernovae nicely compliment those of Type II supernovae by providing the missing ingredient: the iron peak nuclei. Calculations of explosive nucleosynthesis associated with carbon deflagration models of Type I supernovae (Thielemann, Nomoto, and Yokoi 1986) specifically indicate that sufficient iron peak nuclei are produced to account both for the powering of the light curves by radioactive decay of ^{56}Ni and for the mass fraction of iron-group nuclei in galactic matter. Sufficient production of intermediate mass elements like O, Mg, Si, S, and Ca also occurs in deflagration events to be compatible with observations of the spectra at maximum light (Branch 1985), but not to contribute significantly to galactic nucleosynthesis. There remain many questions concerning the nature of the progenitors of Type I supernovae (presumably degenerate core configurations), their evolutionary histories and their frequencies of occurrence in different stellar populations. We also call attention to the fact that the existence of distinctly different sources of iron-peak and intermediate mass nuclei allows for possible differential effects in [X/Fe] for stars in our galaxy.

The heavy elements
The formation of most of the nuclei observed in nature which are more massive than iron occurs primarily by means of neutron capture processes. The abundance pattern in the heavy element region in solar system matter provides clear evidence for contributions from at least two distinct environments characterized by quite different neutron fluxes. The two processes are distinguished by the condition that the characteristic lifetimes against neutron capture are longer than (s-process) or shorter than (r-process) those for beta decay. The neutron capture paths thus defined can differ dramatically as therefore also will the nuclei formed by these reaction sequences. Typically, the r-process capture path lies quite far off the valley of beta stability and gives rise to the production of more neutron-rich isotopes of heavy elements while the s-process forms isotopes which lie on or near the valley of beta stability.

The identification of red giant stars as the site of s-process nucleosynthesis seems firm. The presence of technetium and of enhanced abundances of strontium, yttrium, zirconium, barium, and lanthanum in some classes of red giants supports the view that these elements have been synthesized in the interior and subsequently carried to the surface by convection. Specifically, the operation of the s-process is believed to be associated with the release of neutrons from the ^{22}Ne$(\alpha,n)^{25}$Mg or the ^{13}C$(\alpha,n)^{16}$O reaction proceeding in the thermally pulsing helium shells of red giant stars of intermediate mass ($1 < M < 8$ M$_\odot$) ascending the asymptotic branch. Calculations by Iben and Truran (1978) established that sufficient production of s-process nuclei could occur in this environment to satisfy galactic requirements. Recent reviews of this problem have been provided by Truran (1980), Ulrich (1982), Iben and Renzini (1983), and Mathews and Ward (1985).

The astrophysical site in which r-process nucleosynthesis proceeds remains to be identified (Truran, Cowan, and Cameron 1985). A variety of models have been advanced and explored, ranging from considerations of the expansion of neutron-rich matter from the innermost layers ejected in supernova events yielding neutron star remnants to studies of the conditions realized in the traversal of supernova shock waves through the helium and carbon layers of the supernova envelope. Further discussions and reviews of the nature of the r-process mechanism are provided by Hillebrandt (1978), Truran (1984), and Mathews and Ward (1985). For the purposes of our discussions, we call attention to the facts that the required high neutron densities and corresponding short timescales are certainly strongly suggestive of a violent event and that most models proposed to date are concerned with supernovae environments, and we specifically identify this process with Type II supernovae. As we shall soon discover, this allows for a very straightforward interpretation of the anomalous abundance patterns in the most extreme metal-deficient stars in our galaxy.

Sources of uncertainties in stellar and supernovae yields
A basic measure of the level of uncertainties associated with predictions of stellar yields is provided by the range of values published by different researchers. Scrutiny of the detailed nucleosynthesis predictions for massive stars and associated Type II supernova events published over the past decade (Arnett 1978; Weaver, Zimmerman, and Woosley 1978; Woosley and Weaver 1982; 1986), for example, reveals some substantial differences in relative yields. There are several critical sources of uncertainties in the models and/or input physics which should be noted.

Perhaps the most widely recognized source of uncertainty in the input physics is that associated with the rate of the $^{12}C(\alpha,\gamma)^{16}O$ reaction. The first order abundance effects arising from variations in this rate are obvious: variations in the concentrations both of ^{12}C and ^{16}O and of the relative yields of the products of carbon burning (neon, sodium, magnesium, and aluminum) and oxygen burning (silicon, potassium, sulfur, chlorine, argon, phosphorus, and calcium). These trends are reflected, for example, in the yields determined by Arnett (1978) and by Woosley and Weaver (1982; 1986).

The rate of the $^{12}C(\alpha,\gamma)^{16}O$ reaction also strongly influences the stellar structure and thereby both the overall nucleosynthesis predictions and the character of the resulting remnant. The size of the remnant core and the associated position of the "mass cut" serve to determine particularly the iron yield of a massive star. The core size, in turn, is a function of the entropy structure which depends upon whether the core is convective or radiative. Convective burning provides efficient energy transport and cooling; this favors the transition to a partially degenerate gas, such that the

critical mass for contraction is the Chandrasekhar mass, and relatively small burning cores are realized.   Low ^{12}C concentrations,
alternatively, favor radiative carbon burning cores for which the
Chandrasekhar mass does not play such a dominant role and iron cores
well in excess of the Chandrasekhar limit are subsequently realized
prior to collapse.

Persistent questions concerning the treatment of convection in the cores of massive stars also remain to be addressed.   It
is worth noting that the calculations of Arnett (1978) and Woosley
and Weaver (1986) utilized quite comparable rates for the ^{12}C$(\alpha,\gamma)^{16}$O
reaction, yet their abundance predictions for presupernova cores of
stars of mass M < 15 M$_\odot$ differ drastically.   This would appear to
reflect the level of uncertainties inherent in such studies of
stellar and supernova nucleosynthesis.

One additional source of considerable uncertainty in
nucleosynthesis theory arises from our lack of knowledge of the
nature of the progenitors of Type I supernovae.   Certainly the main
sequence mass range and (presumed) binary evolutionary histories of
Type I progenitors remain to be clearly defined and therefore the
history of Type I activity over the course of galactic evolution is
not known.   Binary models for Type I supernovae are also known to be
extremely sensitive to the assumed rate of mass accretion.   Only over
a relatively restrictive range in mass transfer rates onto degenerate
dwarfs $\dot{M} > 4 \times 10^{-8}$ M$_\odot$ yr^{-1} is it possible for hydrogen and particularly helium burning to proceed with weak shell flashes, insuring
core growth to the Chandrasekhar limit and central carbon ignition.
The ensuing carbon deflagration supernova event (if indeed a deflagration event rather than a detonation event is the result of such
ignition) can then provide ~ 0.5-0.6 M$_\odot$ of iron peak nuclei sufficient to satisfy galactic requirements.   It is expected that such
binary models (or perhaps equivalent single star models) involve
stars of intermediate mass ($\lesssim$ 8-10 M$_\odot$), thus defining their limiting
evolutionary timescales.

Yields from stellar and supernovae nucleosynthesis
In a later section, we will be concerned with the interpretation of the heavy element abundance patterns which are observed
in different populations of stars in our galaxy.   To this end, it
will be convenient to outline a simple prescription for stellar and
supernova nucleosynthesis which reflects the significant trends and
features which have become apparent from our discussions.   We adopt
the following set of assumptions, addressing specifically the yields
of carbon, nitrogen, oxygen, the elements neon-to-titanium, the iron-
peak nuclei, and the s-process and r-process heavy nuclei:

(1)   Intermediate mass stars, $1 \leq M \lesssim$ 8-10 M$_\odot$, represent

the main site of production of ^{12}C, ^{14}N, and the s-
process heavy elements.   These nuclei are formed
during the normal course of evolution of these stars
as luminous red giants.

(2)   Type II supernovae, presumably arising as a conse-
      quence of the evolution of stars in the mass range
      $M > 10\ M_\odot$, are the dominant source of the interme-
      diate mass nuclei neon-to-titanium and of the r-process
      heavy elements. The relative concentrations of ^{12}C and
      iron-peak nuclei in the typical ejecta of such events are
      assumed to be low compared to solar system abundances.

(3)   Type I supernovae are the main contributors to the abun-
      dances of iron group nuclei in the galaxy. While many
      questions remain regarding the possible nature(s) of the
      progenitors of Type I supernovae, it is assumed that they
      typically involve stars of initial main sequence mass
      $\lesssim 8\text{-}10\ M_\odot$.

It is important to note that there thus exists a timescale dependence
of the possible contributions from these diverse sources. The ejecta
of Type II supernovae, the products of massive star evolution, can
begin to contaminate the surrounding interstellar matter on time-
scales $\sim 3\text{x}10^6\text{-}10^8$ years. Alternatively, intermediate mass stars and
Type I supernovae involve stars of mass less than $\sim$ 8-10 $M_\odot$ and
therefore influence the composition of the interstellar medium on
timescales in excess of 10^8 years. It naturally follows that one
might expect the most extreme metal-deficient halo stars present in
our galaxy, presumably formed on a halo collapse timescale $\lesssim 10^8\text{-}10^9$
years, to exhibit abundance patterns which reflect the expected
compositions of the ejecta of the more massive stars.

ABUNDANCE PATTERNS IN METAL DEFICIENT STARS

        The major constraints upon theories of nucleosynthesis
and galactic chemical evolution are those provided by abundance data.
Detailed compilations of the solar system ("cosmic") abundances have
most recently been provided by Anders and Ebihara (1982) and by
Cameron (1982).   Abundance determinations for interstellar matter
provide information concerning the present day composition of the gas
from which new stars are forming. Cosmic ray abundances may provide
a direct probe of the conditions realized in nucleosynthesis
(supernova) environments.   Similarly, observations of abundance
patterns both in evolved stars (e.g. red giants) and in nova,
supernova, and planetary nebula remnants serve as probes of and
constraints upon stellar evolution and the operation of specific
nuclear mechanisms in astrophysical environments.   We are particularly concerned in this paper rather with the abundance history of
the average matter in the galaxy over the course of galactic evolution. We are interested then in the abundances in unevolved stars of
both the disk and halo populations of the galaxy.   Disk population
stars exhibit an overall monotonic variation in the metallicity with
time but no significant relative abundance changes for different
elements.   For metal deficient stars $Z < 0.1\ Z_\odot$, we shall find
deviations from solar abundance patterns which hold important
implications for galactic chemical evolution and nucleosynthesis
theories. We call attention, particularly, to the very nice reviews

of the compositions of field halo stars (Spite and Spite 1985) and of globular clusters stars (Pilachowski, Sneden, and Wallerstein 1983), upon which we have drawn heavily in our subsequent discussions.

        Critical trends in the abundances of carbon, nitrogen, oxygen, the elements neon-to-nickel, and the elements beyond the iron-peak are identified below.

Carbon, nitrogen, and oxygen abundance trends
        Galactic abundance histories for carbon, nitrogen, and oxygen were reviewed by Tinsley (1979). She argued that the high oxygen-to-iron ratios characterizing halo stars were suggestive of the view that they were formed in massive stars (M > 20 M$_\odot$). The observational studies by Sneden, Lambert, and Whitaker (1979) and by Clegg, Lambert, and Tomkin (1981) indeed confirm that oxygen is less deficient than iron in the very metal deficient stars ([O/Fe] ~ $^+$0.5 to + 1.0 for [Fe/H] $\lesssim$ -1.5). This is quite consistent with the numerical results regarding massive star evolution summarized in the previous section. The situation for carbon, alternatively, seems generally compatible with [C/Fe] ~ 0 even for extremely iron-poor stars (although Tomkin and Lambert (1984) have recently determined that carbon over deficiencies relative to iron by factors less than two may exist for a few halo stars). Important recent observations now also provide information on nitrogen in metal deficient stars. Laird (1985) finds both [C/Fe] and [N/Fe] to be approximately constant and compatible with solar in a sample of disk and halo stars which span a range in [Fe/H] of -2.45 to +0.5.

The elements from neon to nickel
        The abundances in field halo stars of many individual elements in the mass region from neon to iron are discussed in detail by Spite and Spite (1985). We are particularly concerned in this paper with the overall abundance trends. Perhaps the most complete and systematic survey of the abundances of the neon to iron group nuclei in metal halo population is that by Luck and Bond (1981; 1985). Their published results for a sample of 21 metal poor stars with Fe/H ratios ranging [Fe/H] = -1.4 to -2.7 reveal several interesting trends. The elements Mg, Si, Ca, and Ti are generally found to be enriched relative to Fe by approximately 0.5 dex, for stars for which [Fe/H] < -2. For the case [Fe/H] $\gtrsim$ -2, the relative abundances of Mg, Si, Ca, Ti, Fe, and Ni are compatible with those of solar system matter. The data also gives evidence for a mild odd-even effect, in the sense that elements containing odd numbers of protons (e.g. Al and Sc) show somewhat greater relative deficiencies at a given [Fe/H] as predicted by explosive nucleosynthesis calculations, but the uncertainties are such that a definite trend is not established.

The heavy elements
        A thorough review of the abundances of heavy nuclei in metal deficient halo stars has again been presented by Spite and Spite (1985; see also Peterson 1976; Spite and Spite 1978; Luck and Bond 1981). The data clearly establishes the existence of depletions in the abundances of the designated s-process elements Sr, Y, Zr, Ba,

La, and Ce relative to iron in stars of low Fe/H; in particular, these trends are evident for [Fe/H] $\lesssim$ -1.5. The existence of systematic differential (or aging?) effects in abundance trends, first emphasized by Pagel (1968) and later reviewed by Tinsley (1979), is substantially confirmed for Sr and Ba. Luck and Bond (1981) note a strong aging trend in Ba, with [Ba/Fe] rising from -1.6 in the most non-deficient giants to approximately zero at [Fe/H] $\approx$ -2.0 to -1.5, while both Y and Zr are found to show similar trends but with lower relative depletions. Tinsley (1979) called attention to the fact that these observed funds in [Y/Fe] and [Ba/Fe] with [Fe/H] are not consistent with the predicted theoretical behavior for such secondary elements (the s-process elements).

One difficulty which arises here concerns ones ability to clearly identify possible modifications in s-process abundance patterns as distinct from r-process abundances. The problem of distinguishing s-process and r-process patterns in stars is complicated by the facts that: (1) most heavy elements receive contributions from both s-process and r-process nucleosynthesis and (2) there exist only a few elements, like europium, which are formed predominantly by the r-process. The critical clue to the nature of these abundance trends was therefore provided by the observation of Spite and Spite (1978) that the abundance of the r-process element europium relative to iron in their sample of metal deficient stars was substantially solar, [Eu/Fe] $\approx$ 0, even for stars of [Fe/H] $\lesssim$ -2.6. Luck and Bond (1985) also determined that the pronounced aging effects observed for Ba and Sr are not evident for the elements Nd, La, Ce, and Pr.

Truran (1981) suggested an alternative interpretation of these abundance trends as being due to the fact that the heavy element abundances in extreme metal-deficient stars are rather products of r-process nucleosynthesis. It is a natural consequence of r-process synthesis that the resulting (Sr-Y-Zr)/Eu and Ba/Eu abundance ratios will be less than those of solar system matter since, in the presence of a high neutron flux, the N = 50 and N = 82 neutron shell closures will be encountered in the neutron-rich regions off the nuclear valley of beta stability, yielding abundance peaks at somewhat lower mass numbers. The heavy element abundance patterns characteristic of the most iron-deficient stars ([Eu/Fe] $\sim$ 0; [Y/Fe] $\sim$ -0.5; [Ba/Fe] $\sim$ -0.8) are therefore entirely consistent with their having an r-process origin. This picture has now been substantially confirmed by the observational results of Sneden and Parthasarathy (1983) and Sneden and Pilachowski (1985); the detailed heavy element abundance patterns they have determined for two extremely metal-deficient stars - HD 122563 and HD 110184, respectively - confirm that the r-process contributions dominate in these objects.

Composition trends in globular cluster stars
It is extremely important to recognize that globular cluster stars can exhibit very similar abundance patterns and trends to those of extreme halo population stars. Detailed analyses for stars in seven clusters have recently been presented by Pilachowski, Sneden, and Wallerstein (1983). Significant observed trends include

the following: (1) high O/Fe ratios are found in a substantial fraction of the studied globular clusters; (2) the intermediate mass elements magnesium, silicon, calcium, and titanium are substantially enriched in abundance relative to iron in most clusters, with values typically of order [(Mg,Si,Ca,Ti)/Fe] ≈ +0.5 for clusters for which the corresponding values of [Fe/H] range from -2.2 (NGC 6397) to -0.9 (NGC 362); and (3) the abundances of the heavy metals Zr, Ba, and La appear to be deficient relative to iron.   Note that all of these trends generally mimic those observed for the metal-deficient halo field stars.   As we shall see, it may be interesting to determine whether these anomalous abundance patterns extend over the entire range of globular cluster Fe/H abundance ratios.

Evidence for a slowly evolving disk metallicity

We have thus far been concerned with possible clues to galactic chemical evolution which are provided by the presence of non-solar abundance patterns in metal-poor stars.   It should be recognized that such trends are characteristic of only a small fraction of the observed stars - those of extreme population II which presumably were formed during the earliest stages of galactic evolution.   Disk population stars, alternatively, may show overall metal deficiencies but typically do not exhibit strong differential abundance depletions of heavy nuclei relative to iron.   Specifically, the age metallicity relation for the solar neighborhood derived by Twarog (1980) indicated that the mean metallicity of the disk increased by a factor ~ 5 over the period from ~ 12 x 10^9 to ~ 5 x 10^9 years ago, and has increased only slightly since then.   In a recent reanalysis of this data, Carlberg et al (1984) find a significantly flatter gradient, with an increase by only a factor ~ 3 over the past 12 billion years.   Realistic models of galactic chemical evolution must of course be successful in reproducing this extremely slowly evolving disk metallicity.   It is also interesting to note that an initial disk enrichment ≈ 30%, such as is implied by the revised age-metallicity relation, would lead to shorter age estimates from studies of cosmochronology.

SOME IMPLICATIONS FOR GALACTIC CHEMICAL EVOLUTION

We wish now to explore some possible consequences of the abundance trends and nucleosynthesis yield predictions surveyed in the previous sections.   We note particularly the following inferences and conclusions:

(1)   The similarities in the abundance patterns characterizing globular cluster stars and extreme halo population field stars are perhaps suggestive of a common nucleosynthesis origin.   It will be of interest to determine whether they were independently contaminated by the ejecta of primarily more massive stars or rather if they were formed from the gaseous debris of a common earlier stellar generation.

(2)   The observed abundances patterns in extreme metal-deficient stars may be understood in a very straightforward manner on the basis

of the simple nucleosynthesis model outlined in an earlier section. The significant abundance trends that have been identified may be summarized as follows: (a) the ratio [O/Fe] is high (~ +0.5) for extreme halo stars, while [C/Fe] $\approx$ 0 and perhaps [N/Fe] $\approx$ 0; (b) values of [Mg/Fe], [Si/Fe], [Ca/Fe], and [Ti/Fe] of order +0.5 characterize stars for which [Fe/H] $\leq$ -2; and (c) the heavy element abundance patterns in extreme halo population stars are consistent with an r-process origin.   Referring to our earlier discussion, we note that the presence both of high ratios of oxygen, magnesium, silicon, calcium, and titanium to iron and of r-process heavy nuclei in extreme halo population stars may simply reflect the contribution of massive stars (M > 10 $M_\odot$) and associated Type II supernovae, which can readily contaminate the gas on a dynamical (halo collapse) time-scale ~ 10^8 years.   These anomalous patterns tend to go away for metallicities in the range [Fe/H] $\geq$ -1.5 as the ejecta of stars of intermediate mass and correspondingly longer lifetimes introduce car-bon, nitrogen, s-process elements, and iron-peak nuclei (from Type I supernovae) into the gas.   It would be interesting to obtain further detailed abundance information for a sample of stars of intermediate population to confirm the nature of these trends.

(3)   The abundance patterns observed in globular cluster stars are consistent with self-enrichment models of globular clusters by means of a small number of massive stars and associated Type II supernovae (see, for example, Fall and Rees 1985).   A ~ 35 $M_\odot$ star, for example, ejects > 10 $M_\odot$ of enriched matter of appropriate compo-sition, hence several such events can in principle contaminate a cluster of mass ~ few x 10^5 $M_\odot$ to a level ~ 10^{-2} $Z_\odot$.   An interesting test of such a self enrichment model might be provided by a careful scrutiny of the abundance patterns of the most metal-rich clusters of Z ~ 10^{-1} $Z_\odot$.   Field stars of such metallicity exhibit patterns of abundances relative to iron which are consistent with those of solar system matter, since the nucleosynthesis contributions from the intermediate mass stars of relatively longer lifetimes have by this time enriched the interstellar medium in carbon, nitrogen, s-process nuclei elements, and iron-peak nuclei.   Self enrichment of a globular cluster must necessarily be realized on a much shorter timescale, hence the abundance patterns of stars in even the more metal-rich clusters should reflect only the contribution from the more massive stars.   If self enrichment indeed occurred for the globular clusters, one might then expect even the metal rich clusters to exhibit, to some degree, both the high O/Fe, Mg/Fe, Si/Fe, Ca/Fe, and Ti/Fe ratios and the r-process heavy element abundance pattern which are found to characterized the most metal-deficient field halo stars.

(4)   The revised age-metallicity relation (Carlberg 1985) may hold potentially interesting implications for nuclear chronology. Thielemann and Truran (1986a,b) find that the results obtained for the galactic-age from the $^{232}Th/^{238}U$ and $^{235}U/^{238}U$ chronometer pairs are strongly dependent upon the degree of initial disk enrichment. The assumption of an initial enrichment at a level ~ 0.3 $Z_\odot$, as implied by the revised age-metallicity relation, yields significantly lower galactic-age estimates than for the case of zero enrichment.

(5)   The revised age-metallicity relation, indicating a very
slowly increasing disk metallicity on a timescale ~ 10-15×10^9
years, may also hold implications for infall. Simple one zone models
of chemical evolution for which the gas density is assumed to fall
exponentially to a level ~ 10 percent on a timescale ~ 10-15×10^9
years generally do not demand infall as a mechanism for maintaining a
present day star formation rate ~ 2 $M_\odot$ yr^{-1} in the disk, since this
level of mass return is naturally provided by the ejecta of dying
stars. The assumption of a relatively constant rate of star formation over the lifetime of the disk, however, would not provide
sufficient return from disk population stars to account for the
current level of star formation activities and some infall may thus
be required.   Future models of galactic chemical evolution must
certainly address this issue.

ACKNOWLEDGEMENTS

One of the authors (J.W.T.) would like to thank the Aspen
Center for Physics for the hospitality afforded him during the period
when much of this paper was written. This research was supported in
part by the National Science Foundation under grant AST 83-14415.

REFERENCES

Anders, E. & Ebihara, M. (1982). Solar system abundances of the
        elements. Geochim. Cosmochim. Acta, 46, 2363-2383.
Arnett, W.D. (1973). Explosive nucleosynthesis in stars. Ann. Rev.
        Astr. Astrophys., 11, 73-94.
Arnett, W.D. (1978). On the bulk yields of nucleosynthesis from
        massive stars. Astrophys. J., 219, 1008-1016.
Arnett, W.D. & Thielemann, F.-K. (1985). Hydrostatic nucleosynthesis.
        I. Core helium and carbon burning. Astrophys. J., 295,
        589-603.
Audouze, J. (1985). C, N, and O isotopes and chemical evolution of
        our galaxy. In Production and Distribution of C, N, O
        Elements, ed. I.J. Danziger, F. Matteucci, and K. Kjar,
        pp. 373-385. Garching: ESO Conference Proceedings No. 21.
Boesgaard, A.M. & Steigman, G. (1985). Big bang nucleosynthesis:
        theories and observations. Ann. Rev. Astron. Astrophys.,
        23, 319-378.
Branch, D. (1985). The optical spectrum of a carbon-deflagration
        supernova. In Nucleosynthesis: Challenges and New
        Developments, ed. W.D. Arnett and J.W. Truran, pp.
        261-271. Chicago: University of Chicago Press.
Cameron, A.G.W. (1982). Elemental nuclidic abundances in the solar
        system. In Essays in Nuclear Astrophysics, ed. C.A.
        Barnes, D.D. Clayton, and D.N. Schramm, pp. 23-43.
        Cambridge: Cambridge University Press.
Carlberg, R.G., Dawson, P.C., Hsu, T. & Vandenberg, D.A. (1985). The
        age-velocity-dispersion relation in the solar neighbor-
        hood. Astrophys. J., 294, 674-681.

Clegg, R.E.S., Lambert, D.L. & Tomkin, J. (1981). Carbon, nitrogen, and oxygen abundances in main-sequence stars. II. 20 F and G stars. Astrophys. J., 250, 262-275.

Diaz, A.I. & Tosi, M. (1986). The origin of nitrogen and the chemical evolution of spiral galaxies. Astron. Astrophys., 158, 60-66.

Fall, S.M. & Rees, M. J. (1985). A theory of the origin of globular clusters. Astrophys. J., 298, 18-26.

Hillebrandt, W. (1978). The rapid neutron-capture process and the synthesis of heavy and neutron-rich isotopes. Space Sci. Rev., 21, 639-702.

Iben, I. Jr. & Renzini, A. (1983). Asymptotic giant branch evolution and beyond. Ann. Rev. Astr. Astrophys., 21, 271-342.

Iben, I. Jr. & Truran, J.W. (1978). On the surface composition of thermally pulsing stars of high luminosity and on the contribution of such stars to the element enrichment of the interstellar medium. Astrophys. J., 220, 980-995.

Lacey, C.G. & Fall, S.M. (1985). Chemical evolution of the galactic disk with radial gas flows. Astrophys. J., 290, 154-170.

Laird, J.B. (1985). Abundances in field dwarf stars. II. Carbon and nitrogen abundances. Astrophys. J., 289, 556-569.

Luck, E.L. & Bond, H.E. (1981). Extremely metal-deficient red giants. II. Chemical abundances in 21 halo giants. Astrophys. J., 244, 919-577.

Luck, E.L. & Bond, H.E. (1985). Extremely metal-deficient red giants. III. Chemical abundance patterns in field halo stars. Astrophys. J., 292, 559-577.

Mathews, G.J. & Ward, R.A. (1985). Neutron capture processes in astrophysics. Rep. Prog. Phys., 48, 1371-1418.

Matteucci, F. (1986). The effect of the new $^{12}C(\alpha,\gamma)^{16}O$ rate on the chemical evolution of the solar neighborhood. Preprint.

Michaud, G., Fontaine, G. & Beaudet, G. (1984). The lithium abundance and constraints on stellar evolution. Astrophys. J., 282, 206-213.

Olive, K.A., Thielemann, F.-K. & Truran, J.W. (1986). Chemical evolution, stellar nucleosynthesis, and a variable star formation rate. Preprint.

Pagel, P.E.J. (1968). Chemical composition of old stars. In Origin and Distribution of the Elements, ed. L.H. Ahvens, pp. 195-204. Oxford: Pergamon Press.

Peterson, R.C. (1976). Constraints on nucleosynthesis imposed by extremely metal-poor stars. Astrophys. J., 206, 800-808.

Pilachowski, C.A., Sneden, C. & Wallerstein, G. (1983). The chemical composition of stars in globular clusters. Astrophys. J. Suppl., 52, 241-287.

Reeves, H. (1974). On the origin of the light elements. Ann. Rev. Astron. Astrophys., 12, 437-469.

Reeves, H., Audouze, J., Fowler, W.A. & Schramm, D.N. (1973). On the origin of light elements. Astrophys. J., 179, 909-930.

Renzini, A. & Voli, M. (1981). Advanced evolutionary stages of intermediate-mass stars. Astron. Astrophys., 94, 175-193.

Sneden, C., Lambert, D.L. & Whitaker, R.W. (1979). The oxygen
          abundance in metal-poor stars. Astrophys. J., 234,
          964-972.
Sneden, C. & Parthasarathy, M. (1983). The r- and s-process nuclei in
          the early history of the galaxy: HD 122563. Astrophys.
          J., 267, 757-778.
Sneden, C. & Pilachowski, C.A. (1985). An extremely metal-poor star
          with r-process overabundances. Astrophys. J. Letters,
          288, L55-L58.
Spite, F. & Spite, M. (1982). Lithium abundances in field halo stars.
          Astron. Astrophys., 115, 357-366.
Spite, M. & Spite, F. (1978). Nucleosynthesis in the galaxy and the
          chemical composition of old halo stars. Astron.
          Astrophys., 67, 23-31.
Spite, M. & Spite, F. (1985). The composition of field halo stars and
          the chemical evolution of the galaxy. Ann. Rev. Astron.
          Astrophys., 23, 225-238.
Thielemann, F.-K. & Arnett, W.D. (1985). Hydrostatic nucleosynthesis.
          II. Core neon to silicon burning and presupernova abun-
          dance yields of massive stars. Astrophys. J., 295,
          604-619.
Thielemann, F.-K., Nomoto, K. & Yokoi, K. (1986). Explosive
          nucleosynthesis in carbon deflagration models of
          Type I supernovae. Astron. Astrophys., 158, 17-33.
Thielemann, F.-K. & Truran, J.W. (1986a). Chronometer studies with
          initial galactic enrichment. In Nucleosynthesis and the
          Implications for Nucleus and Particle Physics, ed. J.
          Audouze and N. Mathieu, pp. 373-388. Dordrecht: Reidel.
Thielemann, F.-K. & Truran, J.W. (1986b). Ages from nuclear chrono-
          logy. In The Extragalactic Distance Scale and Deviations
          from the Hubble Expansion, ed. B. Madore and B. Tully, in
          press. Dordrecht: Reidel.
Tinsley, B.M. (1979). Stellar lifetimes and abundance ratios in
          chemical evolution. Astrophys. J., 229, 1046-1056.
Tinsley, B.M. (1980). Evolution of the stars and gas in galaxies.
          Fund. Cosmic Phys., 5, 287-388.
Tomkin, J. & Lambert, D.L. (1984). Nitrogen abundances in disk and
          halo stars. Astrophys. J., 279, 220-224.
Trimble, V. (1975). The origin and abundances of the chemical
          elements. Rev. Mod. Phys., 47, 877-976.
Truran, J.W. (1973a). Theories of nucleosynthesis. Space Sci. Rev.,
          15, 23-49.
Truran, J.W. (1973b). Nucleosynthesis in red giants. In Red Giant
          Stars, ed. H.R. Johnson, J.P. Mutschlecner, and B.F.
          Peery, pp. 394-433. Bloomington: Indiana University
          Press.
Truran, J.W. (1977). Nucleosynthesis of CNO isotopes. In CNO Isotopes
          in Astrophysics, ed. J. Audouze, pp. 145-154. Dordrecht:
          Reidel.
Truran, J.W. (1980). s-Process nucleosynthesis, stellar abundances,
          and galactic evolution. Nukleonika, 25, 1463-1476.
Truran, J.W. (1981). A new interpretation of the heavy element
          abundances in metal-deficient stars. Astr. Astrophys.,
          97, 391-393.

Truran, J.W. (1983). Nucleosynthesis and the compositions of metal-poor stars. Mem. S.A.Ita., 54, 113-122.
Truran, J.W. (1984a). Nucleosynthesis. Ann. Rev. Nucl. Part. Sci., 34, 53-97.
Truran, J.W. (1984b). Nucleosynthesis constraints on early galactic evolution. In Formation and Evolution of Galaxies and Large Structures in the Universe, ed. J. Audouze and J. Tran Thanh Van, pp. 391-349. Dordrecht: Reidel.
Truran, J.W. (1985). Nucleosynthesis in novae. In Nucleosynthesis: Challenges and New Developments, ed. W.D. Arnett and J.W. Truran, pp. 292-306. Chicago: University of Chicago Press.
Truran, J.W. & Arnett, W.D. (1971). Explosive nucleosynthesis and the composition of metal-poor stars. Astrophys. Space Sci., 11, 430-442.
Truran, J.W. & Cameron, A.G.W. (1971). Evolutionary models of nucleosynthesis in the galaxy. Astrophys. Space Sci., 14, 179-222.
Truran, J.W., Cowan, J.J. & Cameron, A.G.W. (1985). On the site of r-process nucleosynthesis. In Nuclear Astrophysics, ed. W. Hillebrandt, pp. 81-88. Munich: Max Planck Publication MPA 199.
Twarog, B.A. (1980). The chemical evolution of the solar neighborhood. II. The age metallicity relation and the history of star formation in the galactic disk. Astrophys. J., 242, 242-259.
Ulrich, R.K. (1982). The s-process. In Essays in Nuclear Astrophysics, ed. C.A. Barnes, D.D. Clayton, and D.N. Schramm, pp. 301-324. Cambridge: Cambridge University Press.
Weaver, T.A., Zimmerman, G.B. & Woosley, S.E. (1978). Presupernova evolution of massive stars. Astrophys. J., 225, 1021-1029.
Woosley, S.E. & Weaver, T.A. (1982). Nucleosynthesis in two 25 $M_\odot$ stars of different population. In Essays in Nuclear Astrophysics, ed. C.A. Barnes, D.D. Clayton, and D.N. Schramm, pp. 377-399. Cambridge: Cambridge University Press.
Woosley, S.E. & Weaver, T.A. (1986). The physics of supernovae. In Radiation Hydrodynamics in Stars and Compact Objects, ed. D. Mihalas and K.H. Winkler, in press. Dordrecht: Reidel.
Yang, J., Turner, M.S., Steigman, G., Schramm, D.N. & Olive, K.A. (1984). Primoridal nucleosynthesis: a critical comparison of theory and observations. Astrophys. J., 281, 493-511.

DISCUSSION

EDMUNDS: I don't understand what you mean by the massive stars formation giving you initially a <u>high</u> yield.   Surely once you've converted a certain mass of gas to form stars you fixed the "yield" in its classical definition?

TRURAN: If you just plot as a function of time the Z of the material ejected by stars, you have a much higher metallicity at the beginning than you will have on average when you start to see the ejecta from lower mass stars.   So if you are concerned with formation of stars on time scales short compared to the return timescales, as in galactic halo models, a rapid early enrichment does help.

SANDAGE: A comment and a question!   The age of the oldest stars in the Galactic thin disk is known to be the same as the age of 47 Tuc.   Although it is true that NGC 188 forms the lower envelope of field subdwarfs, this does not mean that there are no older field disk stars than NGC 188 ($\sim 10^{10}$ yr) because the envelope in the HR diagram is a function of [Fe/H].   47 Tuc is bluer in its giant and subgiant branch than NGC 188.   It indeed forms the lower envelope of the lower [Fe/H] field subgiants just as NGC 188 does for solar [Fe/H] stars (<u>Ap. J.</u>, 252, 574, 1982) using the [Fe/H] values derived by Helfer (<u>A. J.</u>, 74, 1155, 1969).   His lowest [Fe/H] value for his oldest disk star is $\sim$-0.8 which may be a better starting value in the age-metallicity relation for disk stars at the earliest age.   What, then, would this do to the permissible range of nucleosynthesis ages?

TRURAN: You have two limiting models: One in which you form all the radioactivities at time zero, and another in which you form them continuously.   The time zero model will give 8-10 Gyr, the continuous model will give 15-20 Gyr.   Anything you do to accentuate the early production shortens the timescale.

SANDAGE: So Twarog's curve will give a long timescale.

TRURAN: Yes.   Twarog's curve in that sense will allow for longer ages, systematically.

DEMARQUE: Concerning the lithium 7 abundances of Spite & Spite for halo dwarfs, and their cosmological significance.   Michaud and collaborators have emphasized the importance of diffusion in halo dwarfs from the convective envelope into the radiative layers below.   We have followed Michaud et al's suggestion up and find that indeed, although the temperature at the bottom of the convection zone is never large enough to burn lithium (in agreement with Spite & Spite's conclusion), diffusion could deplete the surface lithium abundance by a factor of three or four.   This abundance should then be viewed as a lower limit in cosmological discussions.

DUNCAN: L. Hobbs and I can show from high S/N spectra that some halo stars indeed show some ^{7}Li depletion.   The Spite's value must be taken as a <u>lower limit</u> to the primordial production.

CAMPBELL: Does Woosley's type II supernova model use population I or population II abundances?

TRURAN: I think he has recently looked also at Population II abundances.

BLAIR: Today when we see a Type II SN, we attribute it to a Population I
star.   Of course the Type II SNe from massive Population II stars (that you
were discussing) would have a different initial composition (i.e., lower heavy
element abundances).   How does the initial composition affect the nucleosynthetic output of Type II SNe and has anyone calculated such models?

TRURAN: From the point of view of the particular groups of nuclei that I have
tried to describe here, I don't think there is a big difference between
Population I and Population II in the results of Woosley's calculations of
supernovae.

KING: In these globular clusters with [Fe/H] less than −2 there must have
been other star deaths and yet we don't see the corresponding nucleosynthetic
products in today's globular cluster stars.   Does that mean that none of that
material could have been retained in those clusters, for some reason?

TRURAN: I like the idea of self-contamination of the clusters, that is: that
you can have a few supernovae going off before the bulk of star formation has
taken place, and then what you have to do is to have them mix the entire
cluster, very efficiently because you have to end up with an extremely narrow
range of metallicity.   Then, the cluster, immediately when it has been
enriched this way, falls into stars, and then it is relatively homogeneous.
Later, any subsequent stellar ejecta leaves the cluster, since there is no
more matter in the way to trap them inside.

KING: But then what about the intermediate abundance clusters which do have
that material?

TRURAN: The total metal enrichment in the cluster is more a matter of how
many of these supernovae can have gone off before the cluster ISM is essentially dispersed.

SANDAGE: Are you in trouble with the self-enrichment of clusters like, say 47
Tuc where $Z \simeq 10^{-3}$, and if you have 10^6 $M_\odot$ of cluster you need 1000 $M_\odot$ of
iron.   Can all that be done by self-enrichment of 47 Tucanae?

TRURAN: But does 47 Tuc also have some range of metallicity?

SANDAGE: No, it's absolutely homogeneous.

TRURAN: Well, then it is harder when you get so high in metallicity.

EDMUNDS: What you are really saying is that while the massive stars are going
off the star formation rate must be zero, otherwise you are bound to see the
low mass stars that formed before and during the enrichment process.

TRURAN: This is in a sense another problem.   The basic point that the chemistry tells us is that the massive stars are the contaminators of the globulars.   How that occurred may be less important than the fact the stars in the
globulars and in the halo had the same nucleosynthetic history.   You can
probably argue reasonably for formation preferentially in the clusters (self-
contamination) or in the background gas.

POPULATION SYNTHESIS IN EARLY TYPE GALAXIES

Robert W. O'Connell
Astronomy Department, University of Virginia
Charlottesville, VA 22903 USA

Abstract.   I review methods of analyzing stellar populations
through evolutionary or optimizing spectral synthesis of their
integrated light. Primary limitations of the techniques appear
to be the problem of accurate abundance determinations and the
inadequacies in our understanding of stellar evolution.   The
best available evidence is that the last epoch of vigorous star
formation in early-type galaxies occurred only 5-10 Gyr ago,
well after globular cluster formation. Observations of high
redshift systems also appear to be consistent with intermediate
epoch star-forming activity in E galaxies. Very little star
formation has taken place since, at least with a normal IMF.

1.   INTRODUCTION

Most of our knowledge of stellar populations is based on
statistical analysis of data for individual stars, particularly in the
form of color-magnitude diagrams.   This is certainly a powerful
technique, whose scope is being continually expanded as more capable
instrumentation is brought to bear.

But only a few stellar systems will ever be within the grasp of
star-by-star analysis. This point is emphasized in Figure 1, which
indicates the range of the Hubble Space Telescope (HST) for color-
magnitude diagram (CMD) work. I have assumed a limiting magnitude of V
= 27 and have ignored image crowding difficulties. The HST can readily
explore young stellar population CMD's to distances of 50-100 Mpc. But
for the kinds of problems I will discuss in this paper, where the main
sequence turnoff is of central importance, even the HST is limited to
the vicinity of the Local Group.

We are therefore clearly obliged to learn how to study stellar
populations through their integrated light. Experience has shown that
this is a less precise and more difficult enterprise than star-by-star
studies, but it also has great promise because it can be applied in
principle to any stellar system in the observable universe. In this
paper I want to survey the strengths and weaknesses of integrated light
techniques by reviewing progress in analyzing the spectra of early type
galaxies (where I include the M31 bulge as well as E and S0 galaxies).

How much can be learned about early-type galaxy populations from
integrated light? Figure 2 shows a state of the art spectrum of M32
obtained by Rose (1985). There is certainly plenty of information
there, comparable to that available for stellar spectral classification

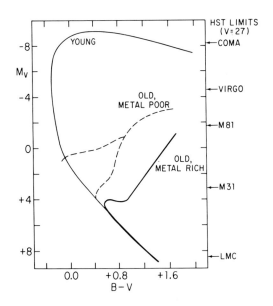

FIGURE 1:   A color-magnitude diagram for several population types with limiting magnitudes for the Hubble Space Telescope indicated at the distances of selected objects.

in the MKK atlas.   The problem facing us is then to determine how many independent population parameters affect the integrated light and how to extract them.

This problem was first confronted by Whipple (1935), who developed techniques to synthesize absorption line strengths and profiles in galaxy spectra. He showed how the compositeness of integrated spectra-- i.e. the   fact   that   different   types   of   stars   make   comparable contributions at a given wavelength and that this mixture changes with wavelength--gave rise to apparent anomalies such as their dwarf-like character and the redness of galaxies compared to stars of the same mean spectral type.   Compositeness   is   a   fundamental   characteristic   of integrated spectra, and meaningful results cannot be obtained from them unless it is properly modeled.   The light from even a simple stellar population is, of course, composite, being a combination of spectra of different temperatures and gravities.

A galaxy, as a superposition of populations with a significant range of individual properties, is yet more composite. I will consider an individual stellar population to be a homogeneous set of stars formed in a short period of time and characterized by their age, t, and metal abundance, Z.   Many other parameters (e.g. helium abundance, mass function, ratio of light to heavy metals, rotation, kinematics, etc), of course, affect the character of populations.   I will mention some in passing, but I want to focus on what one can learn about the gross distribution over t and Z in early type galaxies.

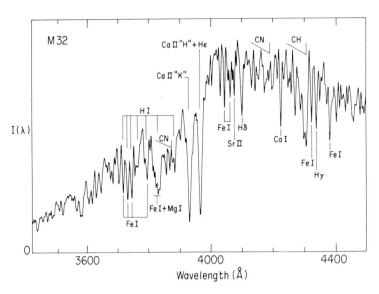

FIGURE 2:   A spectrum of the low luminosity E galaxy M32 with 2.5 A
resolution taken by Rose (1985).   The instrumental intensity scale has
not been converted to flux. All of the high frequency structure in this
spectrum is real, and some of the stronger absorption features are
identified.

2. POPULATIONS IN EARLY TYPE GALAXIES

Compared to spiral galaxies, luminous early-type systems have
relatively homogeneous color and spectral properties (e.g. Lasker 1970,
Faber 1977, Sandage and Visvanathan 1978), which suggests that their
light is dominated by populations with a relatively small range of t and
Z. Since Baade's (1944) classification of E galaxies as Pop II systems,
globular clusters have been considered to be good starting points for
the study of early type galaxies. The evidence available a decade ago
supported the view that globulars and E galaxies formed a continuum of
systems with a common age (~15 Gyr) and a range of Z. However, more
recent line strength and infrared continuum studies (reviewed in
Burstein et al. 1984 and O'Connell 1986) have pretty thoroughly
demolished the concept of a globular-galaxy continuum. The data has not
been satisfactorily interpreted and does not necessarily imply that the
ages of E galaxies and globulars are different. But the newer results
do remove the strongest basis for believing they are the same, and one
is obliged to seek independence evidence concerning the time scale for
star formation in E galaxies.

The classical picture of E galaxy formation, as articulated in the
1957 Vatican conference and Eggen et al. (1962), involved the rapid
collapse and metal enrichment of an isolated structure early in the
universe. Current fashion tends to emphasize a complex interaction with

the environment, perhaps continuing for some time, in which collapse, infall, stripping, accretion, and mergers may all be important mechanisms (see Dressler 1984 for a recent review). The object emerging from these chaotic and episodic processes is likely to be a complex mixture of subsystems, and perhaps it is better to speak of galaxy "assembly" rather than galaxy "formation". Mergers in particular may produce drastic changes. They may convert disk systems into spheroidal systems and in the process generate intense bursts of star formation such as that observed in Arp 220 (Toomre 1977, Schweizer 1985).

If such environmental effects are as important as the theoreticians lead us to believe, the history of star formation in E galaxies may be punctuated by strong discontinuities, perhaps over an extended period. For reasons I will discuss in Sec. 7, it is difficult to determine from integrated light what happened before the last significant epoch of star formation, and so I will concentrate in this paper on the properties of this last episode. As shorthand I will refer to this as the "last burst", though I don't mean to imply it was necessarily a period of sharply increased star formation. A wide range of functional forms for star formation prior to the "last burst" is probably consistent with the data.

CMD studies of nearby dwarf galaxies (reviewed by Aaronson at this conference) provide evidence of the kind of complex star forming histories described above. Many systems once thought to have been quiescent for nearly 15 Gyr are now found to have experienced "last bursts" 3–8 Gyr ago. Detailed CMD studies of the outer bulge of our Galaxy (Whitford 1985, Frogel 1986) and of M31 (reviewed by Mould at this conference) are now technically feasible, providing the opportunity for a combined star-by-star and integrated light attack on these important fiducial systems. Interpretation of the CMD results of the Galactic bulge is controversial, however, and it is clear that good metallicity determinations are essential before the history of star formation can be assessed. The results of Wirth and Shaw (1983) that spiral bulges exhibit larger color gradients than E galaxies also indicate that one must be cautious about extrapolating studies of the fringes of bulges to the nuclei.

3.   SPECTRAL SYNTHESIS TECHNIQUES

The process of spectral synthesis is conceptually simple: one combines individual spectra to produce a model composite. The elements of synthesis are sketched in Figure 3. There are two different approaches, designated "evolutionary synthesis" and "population synthesis" by Tinsley (1980). In evolutionary synthesis, one predicts the spectral energy distribution (SED) resulting from a given CMD, which in turn has been derived from a particular evolutionary scenario or from empirical data. The technique is reviewed by Tinsley (1980) and Barbaro & Olivi (1986). It is simple mathematically and generally is used to explore only a limited parameter space, typically involving no more than 2–3 parameters. It is especially useful in examining the range of

POPULATION SYNTHESIS

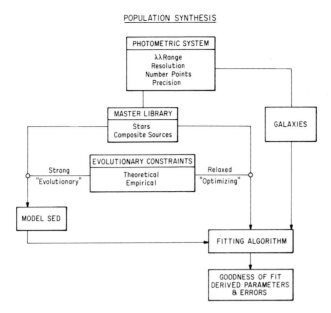

FIGURE 3:   A block diagram showing the main elements of the spectral synthesis technique.

spectral properties which can be induced by variation of the dominant population parameters.

I think a better term than "population synthesis" is "optimized synthesis".   Here, one inverts the evolutionary approach and estimates the most probable CMD responsible for a given observed SED. Historically, this technique began with relatively unconstrained addition of groups of stars in the CMD by trial and error to fit galaxy SED's (e.g. Roberts 1956, de Vaucouleurs & de Vaucouleurs 1959, Baum 1959).   It steadily evolved to the point where a very large volume of parameter space or range of population mixtures may be simultaneously explored by sophisticated automatic optimizing algorithms (see reviews by Pickles 1985 and O'Connell 1986).   The types of linear or quadratic programming methods commonly used appear to have wide utility in astrophysics (e.g. Schwarzschild 1979).

Stellar physics enters both synthesis techniques at the box labeled "evolutionary constraints".   An important distinction is that optimizing techniques explicitly allow relaxation of nominal constraints in order to obtain a best fit to the data. They thus can take direct account of uncertainties in stellar evolution related to convection theory, mass loss, or abundance mixtures, for example, or to the limitations of empirical samples of nearby populations.   Optimizing techniques also allow one to estimate uncertainties in derived parameters and to test their sensitivity to input assumptions.

Optimizing algorithms are usually able to fit composite spectra to the precision expected from observational errors (typically 1-2%) and therefore permit interpretation of the remarkable differences between populations which emerge in high S/N spectra (e.g. Rose 1985). The most pressing need for the optimizing algorithms on the technical side is for more rigorous error and correlation analysis methods.

Spectral synthesis techniques, evolutionary and optimizing alike, are affected by several important limitations:

a. Incomplete understanding of stellar evolution. The approximate nature of our understanding of key phases of evolution becomes painfully obvious when trying to interpret composite spectra. The main source of dispersion in synthesis results by different authors is the latitude allowed by the available empirical and theoretical database in designing evolutionary constraints. One needs to know not only the CMD location of isochrones covering a large range of the relevant parameter space but also the luminosity function along each isochrone, something which has only infrequently been tested empirically. One must also evaluate the importance of certain short-lived or poorly understood phases (e.g. AGB, post-AGB, blue straggler) which by virtue of their brightness or temperature can be significant in integrated light.

Until recently, the best available grid of isochrones was the Yale 77 set (Ciardullo & Demarque 1977), but this required major adjustment for mixing lengths and advanced phases of evolution (e.g. Aaronson et al. 1978, Rabin 1980). VandenBerg (1985) is developing an improved grid which, however, also appears to require significant empirical corrections. It is evident that a better understanding of convection, mass loss, and chemical mixing on the giant branch is essential for synthesis studies as, perhaps, is inclusion of non-zero [CNO/Fe] in model grids (Rood & Crocker 1985).

Optimizing synthesis, by permitting automatic internal adjustment of the models, is less sensitive to this kind of uncertainty than is evolutionary synthesis. Although not widely commented on in the literature, it is commonly found that models which are strongly constrained to follow theoretical isochrones or cluster CMD's, for example, provide significantly poorer fits to early-type galaxy spectra than do relaxed versions (e.g. Pickles 1985). It would be important to examine such discrepancies carefully with a view toward improving both the synthesis modeling techniques and the interiors calculations.

b. Library deficiencies: A good library of stellar SED's (empirical or theoretical) for study of gE galaxies would thoroughly cover a grid of ages of 1-20 Gyr and Z's of 0.1-10 $Z_\odot$. No such library exists or is likely to in the near future. The Galactic disk population near the sun has been well sampled but covers only $Z \sim 0.5-1.5 \ Z_\odot$ for t $\lesssim$ 10 Gyr. Burstein (1985) and Faber et al. (1985) have emphasized that such samples tend to select lower gravity giants than would be appropriate for the study of very old populations. Only partial samples of low metallicity stars are available (e.g. Christensen 1978).

It is possible that few, or no, local analogues exist for some important types of stars in other galaxies. The best evidence for such "missing components" are the overstrong metallic lines in galaxy nuclei (reviewed by Faber 1977) or in M31 globular clusters (Burstein et al. 1984), which seem to require the presence of objects more extreme than the "super metal rich" (SMR) stars in the solar neighborhood. The very metal rich K giants in the Galactic bulge, with [Fe/H] ~ 0.5-1.0 (Whitford and Rich 1983), may be the counterparts of gE giants, but a complete survey of bulge stars will be difficult because of their faintness, image crowding, and the limited low-extinction windows available. It is also essential, though even more difficult, to include SMR dwarfs.

To provide more comprehensive libraries for these and other difficult types, we may well have to rely on model atmosphere calculations. Of course, one can always include the composite spectra of populations (e.g. the M31 metal rich clusters) which themselves may not yet be completely interpretable as elements in libraries to study other systems.

    c.   <u>Interdependence of age and abundance</u>: Effects of age and abundance are difficult to distinguish in integrated light. There is a simple physical explanation for this. The emergent flux from any stellar atmosphere is a superposition of spectra which are characterized by different temperatures, weighted by a function of opacity: $F \sim \int S(T[\tau])e^{-\tau}d\tau$. To first order, increases in opacity caused by a higher abundance, say, are equivalent to a decrease in the mean temperature of the emergent spectrum. Thus, the light of a more metal rich population can be simulated by decreasing the mean temperature (i.e. increasing the age) of a less metal rich population. (A similar opacity-dependent effect operates in stellar interiors.) The following table, based on Rabin's (1980) calculations, illustrates the degree of t/Z interdependence for the integrated broad band colors of models with t ~ 13 Gyr, $Z \sim Z_{\odot}$:

Color	$\partial Color/\partial Log\ Z$	$\partial Color/\partial Log\ t$
U-V	1.27	1.13
B-V	0.37	0.44
V-K	0.68	0.47

The table indicates that a given set of broad band colors for an old, metal rich population can be approximated by a range of (t,Z) models with $\delta \log t \sim -\delta \log Z$ (cf. Figure 4). The clear implication is that age and metallicity must be <u>simultaneously determined</u> using a complete (t,Z) grid of models. It is also evident that data with higher spectral resolution than broad band colors is necessary. As might be expected, the temperature distribution of the population is most strongly constrained by the continuous energy distribution, while the metallicity distribution is most strongly constrained by the line spectrum, though there is enough interaction that these cannot be considered independently (e.g. Pickles 1985). Long wavelength-baseline spectra with resolution of 20-30 A and S/N > 30-50 appear sufficient to separate t and Z effects (cf. O'Connell 1986). The practical difficulty of assembling a proper grid of such spectra is considerable.

The fact that a given stellar SED can be synthesized to first order by combinations of other SED's, implying that the components of a synthesis model do not represent a set of independent basis functions, is responsible for the widely-discussed "nonuniqueness" phenomenon (e.g. the appearance of CMD gaps in loosely-constrained optimizing solutions [Williams 1976]). True astrophysical nonuniqueness is present if three conditions are satisfied: (a) one can find two synthesis models whose fits to the data are statistically indistinguishable; (b) both are astrophysically plausible; and (c) their astrophysical implications are distinct. Ignoring situations where the nonuniqueness is extrinsic to the synthesis technique (e.g. caused by uncertainties in stellar evolution), one can find many published examples satisfying two of these criteria but few meeting all three, at least where adequate S/N is available. I therefore think we have probably been over-concerned with nonuniqueness.

4.   THE LAST BURST OF STAR FORMATION

This section constitutes a non-comprehensive review of what synthesis techniques have enabled us to learn about the properties of the last episode of star formation in early-type systems. Earlier reviews with different emphasis include Faber (1977), Tinsley (1980), and O'Connell (1986).

Broad-band photometry has usually been treated by evolutionary synthesis involving single-generation models. The most definitive results have emerged from near-infrared studies of the continuum, CO, H_2O, and other features (e.g. Aaronson et al. 1978, Frogel et al. 1980, Arnaud & Gilmore 1986) which indicate that the logarithmic slope of the initial mass function (x in the notation of Tinsley 1980) is < 1.5. This implies giant dominance of the IR light and low M/L's and is consistent with a number of higher resolution studies (e.g. Whitford 1977, Cohen 1979, and others cited below).

Unfortunately, as expected from Sec. 3, broad-band colors yield ambiguous results for age and metallicity. Figure 4 illustrates the U-B, B-V loci for the extensive t,Z grid of Rabin (1980), which is based on corrected versions of the Yale 77 isochrones. For $Z < 0.5 \, Z_\odot$, good t,Z separation is possible. But the loci become degenerate for old metal rich populations, and UBVK colors are no less degenerate. Similar ambiguities associated with broad band photometry are discussed by Kennicutt (this conference) in a different context. The nuclei of luminous early-type galaxies fill the "toe" of the dashed outline, and it is evident that the age assigned to populations in this region will

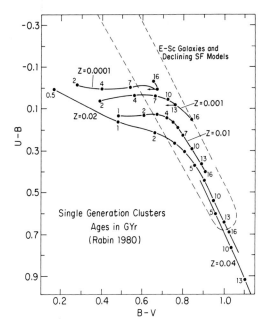

FIGURE 4:   A two-color diagram of evolutionary models for single
generation star clusters of different metallicities, taken from Rabin
(1980).   I have extrapolated Rabin's calculations for solar (Z = 0.02)
and lower values to Z = 0.04.   The arrows show the effect of horizontal
branches in metal poor clusters.   The dashed outline shows the position
of   normal   galaxies   of   all   types   and   of   evolutionary   models   with
exponentially declining star formation rates.   gE nuclei occupy the toe
of the outline; M32 falls at (B-V) = 0.85, (U-B) = 0.41.

be   a   very   strong   function   of   Z.   The   red   envelope   of   nuclear   colors
corresponds   to   15   Gyr   if   Z = Z$_\odot$ but   only   7.5   Gyr   if   Z = 2Z$_\odot$.   However,
the   colors   do   provide   the   interesting   limit   that   Z $>$ 0.5 Z$_\odot$ if   t $<$ 20
Gyr.

        Many   optimizing   synthesis   studies   of   narrow-band   spectrophotometry
(typically   20 A   resolution)   have   followed   the   pioneering   work   of   Spinrad
and   Taylor   (1971).   Most   have   allowed   for   multiple   stellar   generations
and   also   crude   metallicity   mixtures   involving   Galactic   disk   stars   with   Z
~ Z$_\odot$,   SMR   giants,   and   very   metal   poor   globular   clusters.   Age-dating   in
these   studies   is   based   on   determination   of   the   photometric properties   of
the   main   sequence   turnoff,   e.g.   (B-V)$_{msto}$.   The   MSTO   is   well   defined   in
these   models   for   three   reasons: (i)   there   is   a   sharp   break   in   the   main
sequence   luminosity   function   at   that   point;   (ii)   the   density   of   stars
there   physically   determines   the   population   of   the   "downstream"   subgiant
and   giant   branches,   which   provides   a   key   link   between   the   dominant
components   of   a   model;   and   (iii)   the   MSTO   stars   are   the   hottest
significant   components   of   a   given   generation   (at   least   for   metal   rich
compositions).   Depending   on   how   constraints   are   formulated,   (B-V)$_{msto}$

can be determined equally well in single or multiple generation systems. In the latter case it refers to the youngest (bluest) significant generation (the "last burst"). The estimated MSTO color is also only weakly dependent on any mismatch in metallicity between the galaxies and the library, as long as the library spans the colors of the actual turnoff stars.

Figure 5 summarizes turnoff (B-V)'s determined for luminous gE nuclei (or M31's bulge) and for the low luminosity E galaxy M32 in 17 different studies of 11 different narrow-band data sets. For M32 the studies include Spinrad & Taylor (1971), Faber (1972), Pritchet (1977), O'Connell (1980), Peck (1981), Keel (1983), Bruzual (1983), Burstein et al. (1984), and Rocca-Volmerange & Guiderdoni (this conference). The last 3 of these treat 2000-3000 A IUE data. For gE's/M31 the studies include Spinrad & Taylor (1971), Faber (1972), O'Connell (1976), Pritchet (1977), Taylor & Kellman (1978), Peck (1980), Gunn et al. (1981 = GST), and Pickles (1985). An unconstrained optimizing synthesis of broad-band data for M31, including UV photometry from ANS, by Wu et al. (1980) found $(B-V)_{msto}$ < 0.56. It would fall in the bluest M31 bin but is not plotted.

All studies found good evidence for a sharp truncation of the main sequence luminosity function and a reasonably well defined $(B-V)_{msto}$. None found a significant range of turnoffs among normal luminous gE's (though the samples are small) nor a difference between M31's nucleus and gE's. For M32 the various studies are all in agreement that $(B-V)_{msto}$ ~ 0.50 (F7-8 V equivalent spectral type), with a dispersion of only a few 0.01 mags.

The larger scatter in turnoff color for gE's appears to be mainly a product of different assumed evolutionary constraints. For instance, a larger value for the (subgiant + giant):MSTO ratio results in bluer $(B-V)_{msto}$. The study yielding the reddest turnoff color (GST) adopted giant branch loci significantly different from the other studies (cf. O'Connell 1986). The unweighted mean of all the gE/M31 determinations is $\langle(B-V)_{msto}\rangle = 0.65$.

The age corresponding to the turnoff color is a very sensitive function of the population's metallicity. A calibration of the color-Z-age relation is plotted in Figure 5. It is based on the Yale 77 isochrones with a constant temperature shift such that the $Z_{\odot}$ isochrone matches the present sun. For this calibration, $\partial \log t / \partial \log Z \sim -0.65$ at constant $(B-V)_{msto}$.

Synthesis models for M32 are in agreement that solar-abundance libraries yield very good fits. For instance, O'Connell (1980) found a mean residual over 40 wavelengths of only 0.018 mag, comparable to observational error. Individual strong metallic features had residuals which are equivalent to a differential $\langle[X/H]\rangle = -0.02$ with respect to the solar neighborhood library employed, all of whose giants are SMR stars. The plotted calibration read at $Z = Z_{\odot}$ then implies a last burst age of ~5 Gyr for M32.

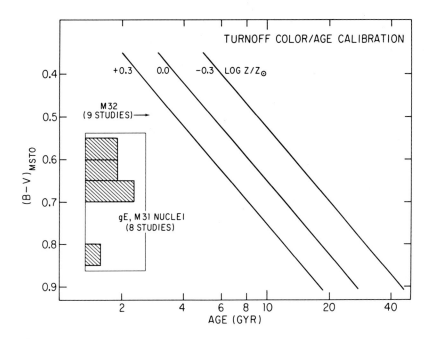

FIGURE 5:   Calibration of the main sequence turnoff color-metallicity-age relation, based on corrected Yale 77 isochrones. The last burst turnoff colors derived from various optimizing synthesis studies of narrow-band spectrophotometry are indicated at the left.   See text for further details.

Metallicity determinations for the more luminous objects are in a less satisfactory state, but various analyses of line strengths indicate $Z \sim 2-3\ Z_\odot$ (see reviews by Faber 1977 and Pagel & Edmunds 1981). Pickles' (1985) study is particularly relevant because it is the only one to apply a grid of model metallicities to a set of galaxies of different luminosity.   He is able to recover the expected Z-luminosity relation (e.g. Pagel & Edmunds 1981), and he finds $Z \sim 2\ Z_\odot$ for gE's. Reading the plotted calibration at this value for the mean gE/M31 turnoff color yields a last burst age of ~8 Gyr.

Rose (1985) has developed an independent approach to this problem. Using very narrow band spectra (2 A resolution), he defines Sr II/Fe I/Hδ indices which are sensitive to the surface gravity (g) mixture of the population but insensitive to metallicity.   He reaches two important conclusions: (i) the g mixtures of metal rich Galactic globular clusters (i.e. 15 Gyr old, moderately metal poor) and E galaxies are different at the 5σ level; and (ii) the light of M32 and gE galaxies is dwarf-dominated at 4000 A, implying significantly brighter and younger turnoffs than in globular clusters.   When Rose's spectra are properly modeled for compositeness it should be possible to obtain good information on [Fe/H] and [X/Fe] as well as turnoff age.

To summarize the available synthesis evidence: metallicities in E galaxies appear to range from $Z_\odot$ to $2Z_\odot$ between M32 and typical gE's. Last burst ages, based on turnoff colors, for the cases studied are 5-10 Gyr, indicating that star formation continued in E galaxies well after globular cluster formation.

5. CRITIQUE

    The conclusions of the preceding section are controversial because of the long-standing supposition that globular clusters and E galaxies are coeval, both completing star formation ~15 Gyr ago. The integrated light studies of both clusters and galaxies constitute a serious challenge to this view. Some of these (e.g. Burstein et al. 1984 or Rose 1985) establish model-independent distinctions between globular clusters and galaxies which cannot obviously be attributed to parameters other than age. The synthesis studies provide no evidence of a dominant 15 Gyr-old component in E galaxies, but these depend on the validity of the modeling process and its calibration. The models are complex, and it is hard to assign uncertainties to derived parameters. It is worth considering potential weaknesses in the three key elements of the argument: the turnoff colors derived, the Z's assumed, and the color-Z-age calibration.

    a. <u>Modeling defects and turnoff color</u>: Considering the wide variety of data sets, modeling techniques, and evolutionary constraints employed, the turnoff color results discussed above are surprisingly robust. Much of the scatter for gE's is due to differing evolutionary constraints, rather than some inherent limitation in the synthesis technique. Various mixtures of theoretical and empirical CMD's are employed by different authors. It would be worthwhile for the community to establish a consensus set of standard CMD's for synthesis studies.

    A number of alternatives to the intermediate age last burst models have been suggested. Some, like blue stragglers, are hard to assess astrophysically and would replace an uncomfortable situation with a bizarre one. However, most plausible alternatives seem unlikely. For instance, observations below 4000 A allow optimizing studies to set good limits on light warmer than the quoted turnoffs. These appear to rule out "combination" models, where an old, cool population and a young (<1 Gyr) hot population combine to simulate an intermediate turnoff. Another plausible alternative, namely the possibility that red horizontal branch stars produce the bluest light, has been ruled out by Rose's (1985) observations.

    Inadequacies in the library will influence the synthesis fits. An attempt to fit a metal rich population (e.g. a gE nucleus) with a library of normal metallicity will overestimate $(B-V)_{msto}$ by several 0.01 mag. because of stronger blanketing in the metal rich stars below 4000 A. Differential blanketing for early K giants as a function of surface gravity (e.g. O'Connell 1973) leads to a similar effect if the library's giants are overluminous with respect to the galaxy's (Burstein 1985). Hence, corrections for known deficiencies in the library are small and would reduce further the gE turnoff colors and ages derived. If M32 were to have slightly subsolar Z, the two effects would tend to cancel.

Thus, there is no obvious reason to suspect serious errors in the derived turnoff colors.

b. <u>The abundance scale and metallicity mixtures</u>: The message of Figs 4 and 5 is that a good estimate of metal abundance is crucial to age determination. There is no more sobering object lesson in this regard than the long debate over the nature of the SMR stars (Spinrad & Taylor 1969, Taylor 1982a & b), whose spectral characteristics are similar to those of gE galaxies. This controversy may be in the process of resolution (Faber et al. 1985) in favor of genuine super-metallicity for these stars. But the uncertainty surrounding the properties of such easily observed objects, where compositeness is not a factor, does not augur well for the problem of galaxy abundances. Evidence that Fe line strengths do not track Mg in galaxies (e.g. Burstein et al. 1984; Davies & Sadler, this conference) further confuses the issue. The long-standing question of non-zero [X/Fe] in galaxies (e.g. Faber 1977) has yet to be resolved, and its implications should certainly be explored. Calculations by Rood & Crocker (1985) indicate that while non-zero [CNO/Fe] has little effect on the giant branch locus, its effect on the main sequence turnoff is not distinguishable from a comparable [Fe/H]. The effects on integrated spectra have yet to be estimated.

The situation is further complicated by the presence of metallicity mixtures in galaxies. The synthesis results discussed above suggest an rms dispersion in log Z of $\sigma < 0.25$. However, Arimoto & Yoshii (1986) have produced evolutionary models including galactic winds which match observed UBV colors if all galaxies are dominated by old (~13–15 Gyr) populations with a large internal dispersion of log Z ($\sigma \sim 0.5$). Although adjustments of these models for corrections to the Yale 77 isochrones and the color calibration of SMR stars will result in a younger mean age and an altered Z distribution (Pickles 1986), the effects of such metallicity mixtures on synthesis results should be examined (cf. Renzini, this conference).

Because of the obstacles to assembling the necessary libraries (cf. Sec. 3), mixed-Z models have commonly combined very metal poor globular clusters ([Fe/H] ~ -1.5) with solar neighborhood disk stars. This type of purely old, mixed metallicity model for gE's (e.g. O'Connell 1976, models D & E) or M32 (e.g. O'Connell 1980, models C & D) fails to fit the data well. Mixtures of modestly metal-poor and metal-rich populations used by Pickles (1985) yield better fits but also indicate intermediate turnoff ages comparable to those of single-population models. Since Pickles' synthesis models reproduce the expected abundance-luminosity correlation, they are evidently reasonably realistic.

The difficulties confronting purely old, large Z-dispersion models stem from the fact that the least metal-rich populations will dominate the blue light, while the most metal-rich will dominate the red light. A large dispersion in Z should therefore lead to a spectrum with significant changes of equivalent metallicity (in both continuum and line strengths) with wavelength. The good UV-to-infrared fits (rms errors of 2-3%) obtainable by models with small Z-dispersion indicate that any such wavelength-dependent effects are small. A significant low metallicity component among the stars near $(B-V)_{msto}$ would be reflected in line weakening below 6000 A, but this is also not observed. Furthermore, the blue continuum limits the contribution of any metal poor population containing AB horizontal branch stars to less than 5% of the V light (cf. Secs. 4 & 6).

In summary, the abundance scale for early-type galaxies certainly requires a good deal of work, and major changes to estimated Z's, and hence turnoff ages, cannot be excluded. However, it appears that very old models including a large metallicity dispersion will have difficulty fitting the data.

c. Isochrone calibrations: In interpreting synthesis results we are at the mercy of the theoretical isochrones and the fact that there is only one fixed point of calibration, namely the sun. The age scale for the corrected Yale 77 isochrones in Fig. 5 is consistent with the earlier calculations of Iben (1967). VandenBerg (1983, 1985) has recently computed models for different values of the convective mixing length parameter, α, and shows that this results in temperature- and abundance-dependent shifts in the isochrones. For α = 1.6, which produces the best fits to globular cluster CMD's, VandenBerg's (1985) calibration matches that in Fig. 5 for log $Z/Z_\odot$ = 0.0 and −0.3 at $(B-V)_{msto}$ = 0.5 but has a significantly smaller $\partial(B-V)/\partial\log t$. This yields older turnoffs at a given color for $(B-V)_{msto} > 0.5$, which would increase the turnoff ages for gE galaxies (though not for M32). Unfortunately, the issue is clouded by the fact that arbitrary corrections to the α = 1.6 isochrones are needed to fit the sun and the observed main sequence for $Z = Z_\odot$, probably owing to difficulties in calculating colors for metal rich compositions.

Changes in the isochrone calibrations seem to be the most promising way of reconciling globular cluster ages and galaxy last burst turnoff ages at present. The targets to aim for would be 15 Gyr isochrones yielding $(B-V)_{msto}$ = 0.50 for $Z = Z_\odot$ and $(B-V)_{msto}$ = 0.65 for $Z = 2Z_\odot$. The VandenBerg recalibration does not, however, shift turnoff colors sufficiently redward. One would have to seek additional important movement in areas (a) and (b) above in order to obtain agreement of the two age scales.

6. STAR FORMATION SUBSEQUENT TO THE LAST BURST

Young stars evidently do not make major contributions to the light of most early-type galaxies, and estimates of post-burst star formation consequently tend to be sensitive to assumptions made about the older population. Most studies of the MS luminosity function in E galaxies (cited in Sec. 4) find a sharp break at the MSTO with an implied e-folding time for the last burst of < several Gyr and that young populations (<1 Gyr) contribute less than 5% of the V light. The conclusion by GST that ongoing star formation may contribute as much as 7% of the V light is a result of the unusual evolutionary constraints adopted for their old populations. The best limits from optical-region data are based on Rose's (1985) Ca II line ratios, which indicate that young populations with a normal IMF contribute < 2% of the V light, implying a current normalized star formation rate of SFR/L_v < 0.004 $M_\odot$ Gyr^{-1} $L_\odot^{-1}$ in M32 and normal gE nuclei.

Vacuum ultraviolet observations allow one to place strong constraints on hot star populations in E/S0 galaxies. For example, the fact that $m_{\lambda1500} - V \equiv -2.5 \log [F_\lambda(1500A)/F_\lambda(5500A)]$ > +2.0 for normal gE nuclei places an upper limit on the contribution of unreddened O stars (which have $m_{\lambda1500} - V = -4.5$) of 0.3% of the V light. All normal E/S0 galaxies studied to date exhibit an excess (UVX) for λ < 2000 A, in the sense that some component hotter than their MSTO's makes significant contributions there (Code & Welch 1979, Oke et al. 1981). International Ultraviolet Explorer (IUE) spectroscopy indicates that the UVX component has T_e > 20,000 K and is positively correlated with metal abundance as derived from Mg line strengths (Faber 1983, Burstein et al. 1986). Note that this correlation for far-UV "blueness" is contrary to that prevailing at longer wavelengths, where metal richness implies redder spectra.

The UVX component has so far eluded positive identification. A number of possibilities have been examined (e.g. Nesci & Perola 1985), but the most likely alternatives are continuing star formation (CSF) with a normal IMF or old, post-asymptotic giant branch (P-AGB) stars evolving rapidly toward the white dwarf cooling sequence. CSF models for the UVX are generally consistent with the optical limits discussed above (Burstein et al. 1986). As stressed in these studies, however, it is difficult to distinguish between the alternatives using only spectral data of the quality available.

Nonetheless, the balance of the available evidence appears to favor P-AGB stars as the UVX component: (i) The spatial distribution of the UVX light is identical with that of the normal, old stellar population (Oke et al. 1981, Bohlin et al. 1985, O'Connell et al. 1986), suggesting a common origin. There is no evidence for the clumping, or possibly disk formation, one expects if massive star formation were present. (ii) The absence of C IV and Si IV absorption lines in high S/N far UV spectra of the nucleus of M31 is inconsistent with a massive star population (Welch 1982) (iii) Finally, UV imagery of the bulge of

M31 reveals no evidence for recent star formation (Bohlin et al. 1985). A CSF component sufficient to produce the UVX light in M31's bulge would include 200 stars earlier than B0.5 V (M_v = -3.6, m ~ 15 $m_\odot$) in the inner 4' of the bulge. Figure 6 shows that none are present; the only UV-bright star in this region has UV colors consistent with a late-B star, and it could be a foreground subdwarf.

There are several interesting ramifications of the P-AGB interpretation of the UVX component. The first concerns the fate of gas lost by stars as they ascend the giant branch, for which the nominal estimate is 0.010 L_v $M_\odot$ Gyr^{-1} with an uncertainty of a factor of perhaps 3 (e.g. Faber & Gallagher 1976). The discovery of hot, X-ray emitting coronae around many normal E galaxies (Forman et al. 1985) apparently eliminates the possibility that this gas is swept away by galactic winds heated by supernovae. Instead, the ejected gas probably cools and produces a steady-state inflow (Nulsen et al. 1984, Sarazin 1986). However, star formation with a normal IMF sufficient to recycle this material seems to be ruled out by the P-AGB interpretation, which implies SFR/L_v < 0.001 $M_\odot$ Gyr^{-1} $L_\odot^{-1}$. (The limit would be even tighter if the cooling flow deposits more gas in the nuclei observed by IUE than is generated there.) Since star formation is the most plausible repository for the recycled material, the implication is that it must take place with an IMF which is not rich in massive stars (e.g. Jura 1977, Fabian et al. 1983, Sarazin & O'Connell 1983). A similar process may take place in the cD galaxies at the center of cluster accretion flows.

Second, the luminosity of P-AGB objects is extraordinarily sensitive to their final core mass after leaving the AGB (Schonberner 1983, Bohlin et al. 1985, Renzini & Buzzoni 1986), with d(log L_{uv})/d(log M_f) < -10. This implies that UVX observations are direct probes of the age, composition, rate of mass loss, and other parameters which can affect M_f in other stellar populations. The sense of the observed UVX-Mg correlation could be produced if the population of more metal rich galaxies were older or experienced higher net mass loss at a given ZAMS mass than more metal poor galaxies.

All of these results refer to E galaxies or spiral bulges. Before closing this section I should note that there is improving evidence for late star formation in S0 disks (Caldwell 1983, Gregg 1985). 50% of the disks in Gregg's sample of 14 objects have detectable populations younger than 5 Gyr, which he interprets as evidence for a mechanism which converts spiral galaxies to S0's.

7. STAR FORMATION PRIOR TO THE LAST BURST

The "last burst" I have been discussing does not necessarily correspond to the "formation time" of galaxies (that is, the first generation of stars) nor need it be a burst (in the sense of a sharp increase in the star formation rate). Rather, it represents the last significant period of star formation. The question of what galaxies

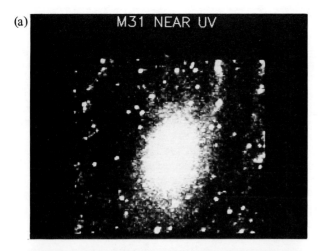

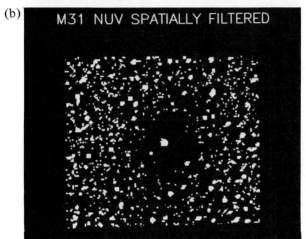

FIGURE 6:   (a) The central bulge of M31 in the middle-UV (1900-2800
A).   The image was obtained by a sounding rocket prototype (15-in
aperture) of the Ultraviolet Imaging Telescope of the Astro mission
(Bohlin et al. 1985).   Total exposure time was 45 sec, and the field of
view is 24 arc-min. on a side. The image is printed at high contrast.
(b) A high contrast, spatially filtered version of the same frame,
where a smoothed image of (a) has been divided into the original to
remove extended structure and isolate stellar images. The many 1-pixel
features which have appeared are photon noise; these are not present in
the center of the frame because of the higher exposure level there. The
nucleus is bright because its characteristic radius is smaller than the
smoothing profile. The filtered image permits detection of faint stars
in the bulge of M31 to $m_{\lambda 2400}$ ~ 17.   If recent star formation were
responsible for the UVX component, over 200 hot main sequence stars
would be visible above this limit in the inner 4 arc-min. Only one real
stellar image is present in this region.

were doing prior to this period was covered extensively at last year's Erice conference (Chiosi & Renzini 1986), so I want to make only a few comments here.

Given the present state of our understanding of stellar evolution, it is difficult to study prior episodes of star formation in integrated light (at least for nearby systems). The problem is illustrated in Figure 7, which sketches how two generations of stars (15 Gyr and 5 Gyr) combine to produce an integrated spectrum. For equal masses in the two generations, the younger will be over 3 times brighter at V (Larson & Tinsley 1978) and even more dominant at shorter wavelengths because of the cool turnoff of the older generation. A synthesis model based purely on the "last burst" young generation would be deficient in stars of redder (B-V)'s as indicated in the figure and would not provide a good fit to the integrated spectrum. However, the uncertainties in evolutionary constraints (cf. Sec. 3 and Renzini & Buzzoni [1986]) are typically 20-30%. If the younger model CMD were adjusted for these (e.g. by increasing the density of stars on the giant branch), evidence for the older generation could be greatly diminished. Probing the early history of galaxies with recent last bursts therefore requires not only high S/N data but also very precise knowledge of stellar evolution. The existing data for nearby galaxies would probably permit a range of functional forms for star formation prior to the last burst, and any inferences concerning that period are likely to be strongly model-dependent.

A better approach to the earlier evolution of nearby galaxies is therefore probably a CMD study of the structure of the giant branch, but, as emphasized above, it is vital that age and abundance effects be separated with confidence.

A large amount of information is now becoming available on galaxies at redshifts high enough to show evolutionary effects directly, and evolutionary synthesis models are being developed to interpret this. In the context of the local evidence discussed above for intermediate age populations in E/S0 galaxies, several points are worth making.

a.   Broad-band color evolution in a quiescent population occurs mainly in the first 1 Gyr after formation. (By "quiescent", I mean a situation where no star formation occurs.) Models by Larson & Tinsley (1978), Bruzual (1983), Hamilton (1985) and Wyse (1985) show that integrated (B-V) reaches ~0.80 only 2-3 Gyr after star formation ceases. This is only ~0.15 mag bluer than for a 15 Gyr-old system. The effects of metallicity must also be considered, especially insofar as they influence the short wavelength spectrum in the restframe. Recent evidence on abundance gradients (e.g. Wirth & Shaw 1983; Efstathiou & Gorgas 1985; Baum et al. 1986; Davies & Sadler, this conference) indicates that large aperture colors for gE models should be adjusted (reddened) for higher than solar abundances. This significantly changes the expected color-redshift relations (Wyse 1985).   Smith & Cornett's (1982) data also indicates that the m_{2400}-V colors of S0 galaxies have a steep dependence on luminosity which probably differs from that for E

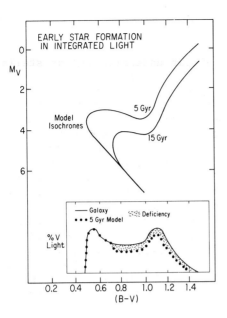

FIGURE 7: Color magnitude diagram for a model galaxy containing two generations. The insert at the bottom shows the distribution of contributing stars over (B-V). A model based only on the younger component will not provide enough red stars, but this deficiency (shaded) is easily concealed by 20-30% uncertainties in stellar isochrones.

galaxies. Because the 2000-3500 A spectrum is so sensitive to age and abundance and is redshifted into the visible region for z ~ 1, careful calibration of all such effects is essential.

b.   A corollary to (a) is that significant deviation from the quiescent colors implies appreciable star formation in the preceding few Gyrs (in the absence of other effects such as nonthermal radiation).

c.   The Bruzual (1983) "μ" models are commonly used as standards of comparison.   In these, primordial gas is converted into stars at an exponentially declining rate. However, it is also assumed that all mass lost by stars on the giant branch is recycled into new stellar populations.   This "second generation" eventually dominates residual "first generation" star formation and can have a strong influence on the short-wavelength spectrum of the model.   Observed trends of color or magnitude with redshift which follow the predictions of the μ models therefore can imply significant recent star formation.

d.   Color differences between star-forming and quiescent systems are not necessarily large.   For instance, the difference between a galaxy which has been dormant for 10 Gyr and one which has been continuously forming stars with a normal IMF during that time is only $\delta(B-V) = 0.4$ and $\delta(U-B) = 0.7$ (Larson & Tinsley 1978). These shifts are comparable to the photometric precision of existing observations of high redshift galaxies. Metallicity differences and dust (which is likely to be associated with episodes of star formation) will further confuse the issue.

Evidence concerning color evolution of high redshift systems presumed to be E/S0 galaxies is reviewed in Chiosi & Renzini (1986), Spinrad (1986), and Oemler (this conference). There is a well defined red envelope which corresponds to old, quiescent systems (e.g. Hamilton 1985). However, starting at $z \sim 0.5$, significant blueward scatter, consistent with the $\mu \sim 0.5$ models, is encountered. This is indicative of reasonably vigorous star formation. Most of the blue systems were selected on the basis of strong radio emission, and their colors are evidently related to spatially-extended bursts of star formation (e.g. Lilly & Longair 1984). However, normal gE's also participate in the blue population (e.g. Eisenhardt 1986, Thompson 1986). Furthermore, recent evolutionary models by Wyse (1985) and Bruzual (1986, Fig. 6) demonstrate that objects constituting the red color envelope for $z < 0.5$ need not be older than 6-10 Gyr at the present epoch and that careful account of metallicity effects must be taken before ages are assigned.

The combined evidence on high reshift objects appears to me to be consistent with the local evidence for intermediate epoch star formation in E/S0 galaxies and with the picture of stochastic galaxy assembly over a long period described in Sec. 2. While the ancestors of local systems would appear as active star-formers at $z < 1$, their morphologies would not necessarily resemble current-epoch E/S0 systems, especially if mergers of spirals commonly produce E's. Shrouding by dust generates a selection effect against detecting systems during their most active star forming phases. Following the last vigorous period of star formation, which we detect locally as the "last burst", a galaxy will quickly evolve (in ~2-3 Gyr) to a position near the quiescent red envelope in the color-redshift diagrams.

8. CONCLUSION

I have been able to mention only a few of the hundreds of integrated light population studies which have appeared in the last 15 years. The synthesis techniques have gradually been refined to the point where they are becoming powerful astrophysical probes of stellar populations. The avenues for further development seem fairly clear:

a. We need a much improved grid of libraries and stellar evolution models covering the relevant Z,t space. Interiors calculations evidently need to be carefully iterated in the context of the available observations. We may not be far from the point when the synthesis studies of galaxies act as important constraints on stellar evolution theory, rather than vice versa.

b.   The abundance scale for galaxies must be established with confidence.   The best subjects for near-term effort would appear to be Galactic disk SMR stars and the extreme SMR stars of the Galactic bulge.   The spatial distribution of abundance in E/S0 galaxies must be mapped by obtaining proper synthesis models for off-nuclear observations.   This requires long wavelength-baseline continuum fluxes as well as line strengths.

c.   The performance of synthesis techniques must be validated by analysis of systems where independent CMD information is available.   For technical reasons, this is difficult with most existing instruments. Christensen's (1972) success at re-deriving the known CMD's of globular clusters was very encouraging, but much more extensive tests are desirable.

d.   Finally, the vacuum ultraviolet is of great importance.   Not only are interpretations of high redshift systems difficult without a thorough understanding of the behavior of this region in local E/S0 galaxies, but the UVX component has important astrophysical implications.   Further, the 2000-3500 A region allows the main sequence turnoff to be studied in isolation from the giant and subgiant branches, which are of comparable importance in the visible region, and offers the best hope for conclusive resolution of the questions surrounding the properties of the last burst.

REFERENCES

Aaronson, M., Cohen, J.G., Mould, J., & Malkan, M. 1978., Ap.J., 223, 824.
Arnaud, K.A., & Gilmore, G. 1986. M.N.R.A.S. 220, 759.
Arimoto, N., & Yoshii, Y., 1986. In Spectral Evolution of Galaxies, eds C. Chiosi & A. Renzini (Dordrecht: Reidel), p. 309.
Baade, W. 1944. Ap.J., 100, 137.
Barbaro, G. & Olivi, F.M. 1986. In Spectral Evolution of Galaxies, eds. C. Chiosi & A. Renzini (Dordrecht: Reidel), p. 283.
Baum, W.A. 1959. P.A.S.P., 71, 106.
Baum, W.A., Thomsen, B., & Morgan, B.L. 1986, Ap.J., 301, 83.
Bohlin, R.C., Cornett, R.H., Hill, J.K., Hill, R.S., O'Connell, R.W., & Stecher, T.P. 1985, Ap.J. (Letters), 298, L37.
Bruzual, A.G. 1983. Ap.J., 273, 105.
----- 1986. In Spectral Evolution of Galaxies, eds. C. Chisoi & A. Renzini (Dordrecht: Reidel), p. 263.
Burstein, D. 1985. P.A.S.P., 97, 89.
Burstein, D., Bertola, F., Faber, S.M., & Lauer, T. 1986, in preparation.
Burstein, D., Faber, S.M., Gaskell, C.M., & Krumm, N. 1984. Ap.J., 287, 586.

Caldwell, N. 1983. Ap.J., 268, 90.
Chiosi, C. & Renzini, A. 1986. Spectral Evolution of Galaxies
        (Dordrecht: Reidel).
Christensen, C.G. 1972. PhD Thesis, California Institute of Technology.
----- 1978. A.J., 83, 244.
Ciardullo, R.B., & Demarque, P. 1977, Trans. Astr. Obs. Yale Univ., 35.
----- 1978. In The HR Diagram, IAU Symposium 80, eds. A.G. Davis
        Philip and D.S. Hayes (Dordrecht: Reidel), p. 345.
Code, A.D., & Welch, G.A. 1979. Ap.J., 228, 95.
Cohen, J.G. 1979. Ap.J. 228, 405.
de Vaucouleurs, G., & de Vaucouleurs, A. 1959. P.A.S.P., 71, 83.
Dressler, A. 1984. Ann. Rev. Astr. Ap., 22, 185.
Efstathiou, G., & Gorgas, J. 1985. M.N.R.A.S. 215, 37P.
Eisenhardt, P.R.M. 1986. In Spectral Evolution of Galaxies, eds. C.
        Chiosi & A. Renzini (Dordrecht: Reidel), p. 403.
Eggen, O.J., Lynden-Bell, D., & Sandage, A. 1962. Ap.J., 136, 748.
Faber, S.M. 1972. Astr. & Ap., 28, 109.
----- 1977. In Evolution of Galaxies and Stellar Populations, eds. B.M.
        Tinsley & R.B. Larson (New Haven: Yale Univ. Obs.), p. 157.
----- 1983. Highlights Astronomy, 6, 165.
Faber, S.M., Friel, E.D., Burstein, D., & Gaskell, C.M. 1985. Ap.J.
        Suppl., 57, 711.
Faber, S.M., & Gallagher, J.S. 1976. Ap.J., 204, 365.
Fabian, A.C., Nulsen, P.E., & Canizares, C.R. 1982. M.N.R.A.S., 201,
        933.
Forman, W., Jones, C., & Tucker, W. 1985. Ap.J., 293, 102.
Frogel, J.A. 1986. In Spectral Evolution of Galaxies, eds. C. Chiosi &
        A. Renzini (Dordrecht: Reidel), p. 143.
Frogel, J.A., Persson, S.E., and Cohen, J.G. 1980. Ap.J., 240, 785.
Gregg, M. 1985. PhD Thesis, Yale University.
Gunn, J.E., Stryker, L.L., & Tinsley, B.M. 1981. Ap.J. 249, 48.
Hamilton, D. 1985. Ap.J., 297, 371.
Iben, I. 1967. Ap.J. 147, 624.
Jura, M. 1977. Ap.J., 212, 634.
Keel, W.C. 1983. Ap.J., 269, 466.
Larson, R.B., & Tinsley, B.M. 1978. Ap.J., 219, 46.
Lasker, B.M. 1970. A.J., 75, 21.
Lilly, S.J., & Longair, M.S. 1984. M.N.R.A.S., 211, 833.
Nesci, R., & Perola, G.C. 1985, Astr. & Ap., 145, 296.
Nulsen, P.E., Stewart, G.C., & Fabian, A.C. 1984. M.N.R.A.S., 208, 185.
O'Connell, R.W. 1973, A.J., 78, 1074.
----- 1976. Ap.J., 206, 370.
----- 1980. Ap.J., 236, 430.
----- 1986. In Spectral Evolution of Galaxies, eds. C. Chiosi &
        A. Renzini (Dordrecht: Reidel), p. 321.
O'Connell, R.W., Thuan, T.X. & Puschell, J.J. 1986, Ap.J. (Letters),
        303, L37.
Oke, J.B., Bertola, F., & Capaccioli, M. 1981. Ap.J., 243, 453.
Pagel, B.E.J., & Edmunds, M.G. 1981. Ann. Rev. Astr. Ap., 19, 77.
Peck, M. 1980. Ap.J. 238, 79.
----- 1981. Private communication.

Pickles, A.J. 1985. Ap.J., 296, 340.
----- 1986. In Structure and Dynamics of Elliptical Galaxies, IAU
      Symposium 127, ed. T. de Zeeuw (Dordrecht: Reidel), in press.
Pritchet, C. 1977. Ap.J. Supp., 35, 397.
Rabin, D.M. 1980. PhD Thesis, California Institute of Technology.
Renzini, A., & Buzzoni, A. 1986. In The Spectral Evolution of Galaxies,
      eds. C. Chiosi & A. Renzini (Dordrecht: Reidel), p. 195.
Roberts, M.S. 1956. A.J., 61, 195.
Rood, R.T., and Crocker, D.A. 1985. In Proc. ESO Workshop on Production
      and Distribution of the CNO Elements, eds. I.J. Danziger, F.
      Matteucci, & K. Kjar (Garching: ESO), p. 61.
Rose, J.A. 1985. A.J., 90, 1927.
Sandage, A. & Visvanathan, N. 1978. Ap.J., 225, 742.
Sarazin, C.L. 1986. In Proc. Green Bank Workshop on Gaseous Haloes
      Around Galaxies, eds. J. Bregman & F. Lockman, in press.
Sarazin, C.L., & O'Connell, R.W. 1983. Ap.J., 268, 552.
Schonberner, D. 1983. Ap.J., 272, 708.
Schwarzschild, M. 1979. Ap.J., 232, 236.
Schweizer, F. 1986. Science, 231, 227.
Smith, A.M., and Cornett, R.H. 1982. Ap.J., 261, 1.
Spinrad, H. 1986. P.A.S.P., 98, 269.
Spinrad, H. & Taylor, B.J. 1969. Ap.J., 157, 1279.
----- 1971. Ap.J. Supp., 22, 445.
Strom, K.M., & Strom, S.E. 1978. A.J., 83, 73.
Taylor, B.J. 1982a. Vistas in Astronomy, 26, 253.
----- 1982b. Ibid, p. 285.
Taylor, B.J., & Kellman, S.A. 1978. Ap.J. Supp., 37, 101.
Thompson, L.A. 1986. Ap.J., 300, 639.
Tinsley, B.M. 1980. Fund. Cosmic Phys., 5, 287.
Toomre, A. 1977. In The Evolution of Galaxies and Stellar Populations,
      eds. B.M. Tinsley & R.B. Larson (New Haven: Yale Univ. Obs.),
      p. 401.
VandenBerg, D.A. 1983. Ap.J. Supp., 51, 29.
----- 1985. Ap.J. Supp., 58, 711.
Welch, G.A. 1982. Ap.J. 259, 77.
Whipple, F.L. 1935. Harvard College Obs. Circular 404.
Whitford, A.E. 1977. Ap.J., 211, 527.
----- 1985. P.A.S.P., 97, 205.
Whitford, A.E., and Rich, R.M. 1983. Ap.J., 274, 723.
Williams, T.B. 1976. Ap.J., 209, 716.
Wirth, A., and Shaw, R. 1983. A.J., 88, 171.
Wu, C.C., Faber, S.M., Gallagher, J.S., Peck, M., and Tinsley, B.M.
      1980. Ap.J., 237, 290.
Wyse, R.F.G. 1985. Ap.J., 299, 593.

DISCUSSION

WHITFORD: Bob has alluded to the advantages of the star-by-star approach for stellar systems that are near enough. The color-magnitude diagram of the Galactic bulge population in the -8° window has been observed by Donald Terndrup of Santa Cruz.   The CTIO CCD photometry reaches almost 2 mag deeper than the turnoff. The foreground disk stars are distributed on the diagram as expected from the Bahcall-Soneira model.   The late M giants (AGB stars) identified by Blanco stand out because of their very red V-I colors.

When Don VandenBerg's theoretical isochrones for solar metallicity are superposed, the mean trend line through the turnoff stars suggests an age since formation of 10-12 Gyr.   There is an unavoidable scatter in magnitude caused by front-to-back distance dispersion, scatter in V-I color caused by metallicity spread, and possibly scatter along a diagonal line caused by age spread.   The diagram does not support the view that the turnoff stars are predominately those evolving from a major burst of star formation as recent as 5-8 Gyr.   This evidence may become somewhat more marginal when isochrones for a population of twice solar metallicity become available.

O'CONNELL: This and the CMD work discussed by Mould on the M31 bulge are very important lines of attack on the problem of early type systems.   I have two generalized concerns: First that it is, as you emphasized, essential to determine the metallicities of these stars since a factor of two increase in Z will decrease the age estimated from the turnoff by 40%.   Second, that Wirth found color gradients in the bulges of spirals were much steeper than in elliptical galaxies, perhaps indicating that the outer parts of spiral bulges, where it is possible to obtain CMD's, are not entirely representative of the interiors or of E's.

AARONSON: It may not be all that surprising that residual star formation has gone on for a great length of time in the nucleus of ellipticals.   The really interesting question is how globally this phenomenon is distributed over galaxy volumes.   Do you have any data (i.e. off-nuclear spectra) which might address this point?

O'CONNELL: Yes, galaxy formation models like Larson's do, as you say, tend to have continuing star formation in the nucleus for long periods.   I have very little off-nuclear data.   A number of broad-band photometry studies (e.g. Strom & Strom 1978), however, have found only mild color gradients in E galaxies, suggesting that the nuclear population as observed in, say, 1 kpc apertures, is not atypical of much of the galaxy's light.   Recent work has shown that while the metal abundance declines outward, it remains above solar to relatively large radial distances.

AARONSON: I don't think the general lack of color gradients in ellipticals tells you much, since as you pointed out, the coupling of age and abundance makes broad band color insensitive. For instance, a decreasing metallicity combined with an increasing age could keep the color relatively fixed as you move out from the nuclear regions.

O'CONNELL: I certainly agree that there are physically plausible combinations of age and metallicity gradients which would tend to suppress broad band color variations.   The problem of separating these effects is, of course, very tricky, especially with the lower S/N data available at large radii.   It deserves a careful treatment, and no one has done that yet.   But your original question asked what fraction of a galaxy's total light might originate in an intermediate age population, and my answer is that, based on what we know now, it could be a large fraction.   Metal abundance is a key factor in estimating ages, and the higher is Z, the lower is the age estimate.   Recent Mg line strength studies by Baum, Efstathiou, Davies & Sadler, and Faber & Burstein are quickly improving the available database.   All these tend to show supersolar abundances to large radii, though the Z calibration is not very secure.

DAVIES: The metallicity gradients indicated by Mg_2 are from twice solar to solar for 0.1-1.0 R_e. The metallicity change reflected by the Mg_2 index must be considered uncertain because the Fe features do not track Mg_2. In Galactic K III's, Guinan & Smith (PASP, 1984) and Faber et al. (ApJ Supp. 1985) show that Mg_2 is a sensitive surface gravity indicator.   Thus, at least some fraction of the Mg_2 gradient is likely to be due to a population change.

The Hβ strength is often reduced toward the center due to weak AGN emission.   Correction for this may increase the size of the intermediate age population inferred in the center.

O'CONNELL: Yes, I think we don't understand the behavior of the Fe, Mg, CN, Ca, and other features in early type systems (including globular clusters in M31) very well.   That has to be a high priority problem for the near future.

Concerning Hβ, correction for nuclear Balmer emission would certainly be important in some circumstances.   I should emphasize, however, that the primary hold on the MSTO and hotter starlight comes from the overall continuous energy distribution and the "filling-in" of strong metallic absorption features in the cooler population rather than from the Balmer lines themselves.

BURSTEIN: Sandy Faber, I, and our collaborators have consistently found that while all the Fe, Mg, and CN indices are correlated with each other, the increase in Fe is substantially less than one would predict from Galactic stars.   This fact was known at the Yale Symposium in 1977.   We now have data for over 300 galaxies with the same effect.   As a result, I feel that one must take seriously the notion that Fe and Mg abundances might not be related in other galaxies as in ours.

O'CONNELL: I agree that there is good evidence for this effect (and
I've been trying to get you to say so since 1976!). In my data it shows
most strongly in Ca II and Fe lines which do not track Mg and Na. Since
we seem finally to be making progress on the nature of the SMR stars, I
hope we can also soon better understand the galaxies, which seem to have
similar patterns of line strengths. Of course, the possibility of non-
zero [X/Fe] adds an extra and unwanted dimension to synthesis analysis
of galaxy spectra. Bob Rood is looking into the effects of non-zero
[CNO/Fe] on isochrones, and someone should evaluate the net effect on
integrated spectra. It's possible that this kind of abundance variation
could explain some of the interesting differences now apparent between
the Galaxy and M31 globular clusters and/or galaxies.

DAVIES: What is the evidence that the UV light tracks the old metal
rich light?   Even if this is true, does it require that UV light is P-
AGB?

O'CONNELL: A number of studies based on IUE spectroscopy show that the
UVX component has a radial distribution remarkably similar to, say, the
V or R brightness distribution. On a larger scale (the IUE aperture is
only 20" long), the surface brightnesses of the UIT images I showed of
the M31 bulge are also in good agreement with Kent's R band photometry
out to 100".   None of the UV studies indicates any spatial clumping as
might be expected from recent star formation.

     No, this doesn't require P-AGB light. I'm sure there are plausible
mechanisms to convert inflowing gas into new stars proportional to the
local star density.   However, I think the spectroscopic and imaging
results on M31 are difficulties for normal IMF star formation at least
in that one case.   Old population alternatives to P-AGB light are
unusually hot horizontal branch stars or accreting white dwarfs, for
example; but these all have their own difficulties. One attraction of
the P-AGB stars is that they are direct descendents of the dominant old
population. We know they are there, and, with a little effort, we can
estimate their properties with some confidence.

KING: Would you comment on the small amount of current star formation
in the centers of NGC 205 and NGC 185? We would not see it if they were
outside the Local Group. Could it be general in ellipticals?

O'CONNELL: These are very interesting cases and good candidates for HST
studies. One possibility is that gas has recently been injected through
tidal interactions with M31 or dwarf irregulars.   Another is that
cooling flows in such lower mass systems are able to generate stars with
more normal IMF's than in more massive systems, though I don't know if
these objects show X-ray evidence for such flows or if this is
astrophysicaly plausible.   Concerning similar situations in other
galaxies, I think Burstein and others have shown that the integrated
optical light of systems like NGC 205 shows clear evidence of the young
population, so a comparable component would be detectable even in
distant systems. UV observations would be able to detect young stars to

much lower levels. I should also remark that with integrated spectra of good quality, one would readily be able to distinguish between a minority massive star component like these and an intermediate age turnoff for the old population.

BURSTEIN (to KING): Faber, Bertola, Buson, Lauer, and I have found that the level of far-UV emission ($\lambda < 2000$ A) from normal elliptical galaxies is correlated with optical absorption line strength, as measured by Mg_2. However, this correlation is in the opposite sense of all other such correlations (as found originally by Faber).

ROCCA-VOLMERANGE: We present in the poster session a recent analysis of high resolution spectra of S0/E galactic nuclei obtained with the IUE satellite. Three main features result from this study: (1) From a simple spectral synthesis with observed stellar spectra we find the main stellar population contributor between 2000 and 3000 A is a uniform F8 spectral type, which indicates an age of about 5 Gyr. (2) Some galaxies have no UV excess (or turnup below 2000 A). (3) An interpretation with an evolutionary model gives a normal exponentially decreasing history of the star formation but no burst at 5 Gyr.

O'CONNELL: One of your objects is M32, and I included your result in my diagram of turnoff determinations. As you say, M32 does not show a UV upturn below 2000 A, but the flux measured there does indicate a very small amount of hot OB starlight, something on the order of 0.02% of the V light. So there seems to be a hot component even if it is much smaller than in more metal rich systems. I would be concerned that any determination of the early star formation history is sensitive to the details of the models, but I agree that the M32 data does not require a sharp increase in star formation at 5 Gyr.

ALTNER: From the IUE data on the UV upturn in the ellipticals can you estimate the number of blue horizontal branch stars contributing and if so, is there any evidence for a second parameter effect?

O'CONNELL: There is a fair amount of scatter in the UVX-metallicity correlation found by Burstein et al., but I don't know whether there is good evidence for more than one parameter. Horizontal branch stars are one possible candidate for the UVX component. Their difficulty is that the UV upturn in galaxies is stronger than observed in globular clusters; or, to put it another way, the galaxies require hotter components in the far-UV than found on typical cluster horizontal branches. This was demonstrated quantitatively by Wu et al. (1980) for M31. Of course, mass loss and other parameters affecting the HB could be different enough in galaxy nuclei that the hottest tip of the HB is more strongly populated there. Ciardullo and Demarque (1978) argued that a sufficiently old and metal rich population will populate the blue HB.

RENZINI: In my afternoon talk I will try to defend an "old age" for ellipticals.

ZINNECKER: It was not clear to me what fraction of the total mass of M32 or gE's participate in that last burst of star formation. Presumably a large fraction is much older than that burst. The second point I would like to raise is the question to what extent your results are affected by the large dispersion in metallicity that will undoubtedly be present in gE's.

O'CONNELL: I think it's fair to summarize various people's synthesis results this way: given the present uncertainties in evolutionary isochrones, the data do not <u>require</u> any contribution from stars significantly older than the quoted turnoff ages. In the case of M32, excellent fits were obtained for a population with $Z = Z_\odot$ and a CMD like M67, which VandenBerg would date at 5 Gyr.   If you stretch the evolutionary constraints, you can squeeze older populations in. I used that approach to estimate the 5σ upper limit for an older population to be 50% of the V light. I believe Pickles finds 20-30% of the light to be older in his gE models.

There must, of course, be a finite dispersion in metallicity in galaxies.   But again, the various synthesis studies are unanimous in limiting the amount of very metal poor starlight (e.g. globular clusters like M15) to about 5% of the V light in gE galaxies.   The main difficulty is putting together a stellar library suitable to create a grid of models with different Z's and ages. Pickles has done the best job of this so far.   He was able to derive the expected abundance-luminosity relation directly from the synthesis models, and his turnoff ages do not differ significantly from those of single-metallicity models.   So, I don't think an internal dispersion in Z affects the derived ages very much.

EDMUNDS: The "clumping" of last burst ages might be an artefact.   If the ellipticals enrich themselves in heavy elements as star formation proceeds, what you see now is a superposition of old, metal-poor CM diagrams and younger metal-rich ones.   The effect of metallicity is to force the turnoff point for a given age to lower luminosities, so you may get all the turnoff points in about the same region of the CM diagram, although stars of widely different ages are present.

O'CONNELL: I agree that this effect can certainly occur, which is why it is essential to obtain good metallicity estimates for the turnoff stars. If 50% of the turnoff light were below solar metallicity, say, then well over 50% of the net blue light of the galaxy would be below solar because the metal poor giant branch is hotter and brighter than the more metal rich giant branch.   The fact that the gE line spectra from the UV through the infrared seem to require solar or higher abundance stars seems to rule out any major contribution (over about 10-15%) from metal poor stars.

DJORGOVSKI: The remarks about young ages of ellipticals may be somewhat misleading since it is assumed that the last burst of star formation was practically a delta function.   If one allows for continuous star formation, the ages may be substantially larger.   For example, the very red, evolved ellipticals found by Don Hamilton at $z \sim 0.6-0.8$ seem to be very old.

Also, a comment on Ivan King's question: Galaxies of the NGC 205 and 185 type are probably a completely different kind of object from the "usual" elliptical and should not be compared to them.

O'CONNELL: I hope I was clear about my "last burst" terminology. I mean by that the last significant epoch of star formation. As you suggest, the data do not require that a sharp increase in star formation occurred then, nor do the models necessarily assume that. Little information is available so far from local ellipticals on the history of star formation prior to the last burst. The galaxies could indeed have started forming stars 15 Gyr ago. However, the possibility of earlier star formation does not affect the conclusion that it continued until 5-10 Gyr ago—i.e. it does not change the $(B-V)_{msto}$ or Z estimates.

If the galaxy formation process involves strong interaction with the environment (e.g. mergers, infall), one expects star formation to occur stochastically over a long period. Spectra and colors of high redshift systems seem to be consistent with this sort of picture and with the relatively late "last bursts" found by synthesis in local objects. Once star formation ceases, color evolution to the red envelope you mentioned is quite rapid (2-3 Gyr), so it's not surprising that a number of red ellipticals are found at all redshifts up to $z \sim 1$. If events like mergers are common, the ancestors of local E's at lookbacks of 5-10 Gyr may not resemble current-epoch ellipticals at all.

DJORGOVSKI: Many normal ellipticals show some evidence for accretion, e.g. cooling flows. These are capable of providing a young population, which may contribute a large fraction of the total light atop an otherwise old galaxy.

O'CONNELL: Yes, this is a very interesting possibility. There is increasingly good evidence from my observations and those of Fabian's group that such "accretion populations" are common in cD galaxies with large cooling flows. The best local example is NGC 1275. We have recently found that about 50% of the other accreting cD's in our sample show significant UV excesses probably associated with accretion populations. Abell 1795 is the most prominent. Constraints on the fraction of the cooling flow which is being converted to visible stars and the IMF are not good yet.

With regard to high redshift objects, it would probably be hard to distinguish accretion populations from other kinds of young generations with optical data, though the restframe vacuum UV might hold some clues if accretion IMF's are truncated, as we suspect. It's possible that X-ray observations with AXAF could establish the onset epoch for cooling flows.

UNIDENTIFIED: Do the color-magnitude diagrams for the various windows support a trend toward lower metallicity with increasing distance from the nucleus of the Galaxy?

WHITFORD: Yes. Preliminary reductions of Terndrup's CCD fields corroborate Blanco's well-known result that bulge samples at higher latitudes show a sharp decline in the fraction of the giants in the sample that are classified M6-M9.   On the CMDs these are set apart by their luminosity and very red color. The simplest explanation is a shift in the metallicity distribution function in the higher-latitude samples that diminishes the fraction of the population that is sufficiently SMR to survive mass-loss attrition and become AGB stars.   The trend of the ridge lines of the whole giant branch on three of Terndrup's CMDs (again preliminary reductions) is toward bluer colors for the higher latitude fields, as would be expected for decreasing average metallicity.

THE EVOLUTION OF STELLAR POPULATIONS IN GALAXIES

Augustus Oemler, Jr.
Department of Astronomy
Yale University
New Haven, CT

INTRODUCTION

That galaxies evolve is a trivial statement: galaxies are made of stars, stars evolve, and therefore galaxies evolve. But over and above such minimal evolution of the older stellar content of a galaxy, which is often called *passive evolution*, there exists the possibility of *active* galactic evolution, by which I shall mean evolution of the number and properties of the young stars in a galaxy. Several of the papers in this volume present reasons for believing in active galactic evolution. Aaronson has discussed the color-magnitude diagrams of dwarf spheroidal galaxies in the Local Group, several of which show evidence for an extended history of star formation in what are now quiescent galaxies. Kennicutt has described how the variation in the colors of galaxies along the Hubble Sequence can be explained by a variation in their star formation histories, ranging from constant star formation for the typical Sc to a rapidly decreasing star formation rate (SFR) in early types. O'Connell has discussed the spectral synthesis of E and S0 galaxies, the results of which almost invariably suggest that they, like the dwarf spheroidals, were forming stars quite vigorously until fairly recent times.

Together, these deductions are, I think, quite compelling, but it must be remembered that most are only deductions, and are dependent on a wide range of assumptions. It would, therefore, be preferable to observe directly the evolution of galaxies, and, for once, nature has been kind to us: observations of galaxies at high redshifts provide direct information on their properties at earlier epochs. Furthermore, much information about stellar populations can be obtained from the integrated spectral energy distribution of a galaxy, one of the few properties of these distant objects which can be measured easily from the ground.

Direct observations of galactic evolution are, therefore, possible, but they are not easy. High redshift galaxies are faint, difficult to find and to observe. A more fundamental problem is that we must establish the present-day properties of the objects which we are observing at an earlier epoch, since it is only by the change in their properties that we can determine their evolution. Because of these problems, we have today the best information about some of the least typical galaxies.

It will be helpful to discuss the possible evolution of various classes of galaxies in the context of two models. One is that of passive evolution, which in the limit of objects in which the last epoch of star formation was very long ago, approaches the case of no evolution. Another, which I shall call the *standard model*, is that outlined by Kennicutt. The SFR in a galaxy is assumed to vary with time in a simple manner, for example, like $\exp(-t/\tau)$, where τ is constant for a galaxy and t is its age. Given a present age, metallicity, and initial mass function (IMF), τ uniquely determines the present-day colors. Thus, in this model, the distribution of star formation histories is completely constrained by the distribution of galaxy colors today. The expected evolution of the rest-frame colors of various Hubble types, taken from the calculations of Tinsley (1980), assuming solar metallicity, a Salpeter IMF, and a Hubble constant, $H_0 = 50$ km s^{-1} Mpc^{-1}, is illustrated in Figure 1. The evolution over the last 8 billion years is quite mild for all types. The extremes of early and late types evolve very little: E's evolve only because of the burning-down of the main sequence, Scd's, which have a constant star formation rate, evolve only because of the gradual build-up of an underlying old population. Sab's, with the most rapidly changing stellar populations, evolve the fastest, but even their evolution only amounts to about 0.18 in B-V during the last 8 billion years.

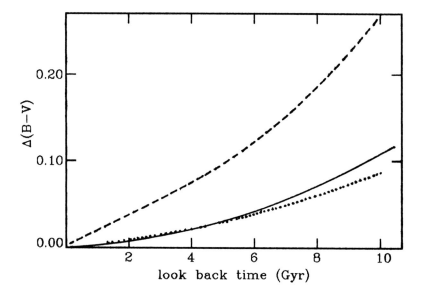

FIGURE 1. Evolution of the rest-frame B-V colors of Elliptical (solid line), Sab (dashed line), and Scd (dotted line) galaxies in the standard model, from calculations by Tinsley (1980).

Our expectations, then, are for very modest changes in the stellar populations in galaxies, at least out to redshifts of the order of one. In the following sections, I shall summarize what is known about the change with redshift of the properties of several classes of objects. After all of the observations have been presented, I shall address the question of whether they can be assembled into a coherent picture of galactic evolution.

GIANT ELLIPTICALS

If we ignored the evidence presented by O'Connell, giant ellipticals would seem the least likely candidates for evolving galaxies. They show, in general, no signs of ongoing star formation; their colors are consistent with those of a very old population (Searle, Sargent, and Bagnuolo 1973); and their lack of disks suggest a history of star formation no more extended than one collapse time ($\sim 10^9$ years). On the other hand, these same characteristics make them particularly easy objects to study. Because they are very red and form a very homogeneous class of objects, it is easy to detect in them the presence of quite small populations of anomalously blue stars. Gregg (1984) has shown that differential comparisons between early-type galaxies can detect young and intermediate-age populations contributing as little as 3 per cent of the visual light of the galaxy. Furthermore, it is easy to identify the high-redshift progenitors of today's giant E's, particularly those that are the brightest members of clusters of galaxies or the hosts of strong radio sources.

Because of these observational advantages, and because of their usefulness in other cosmological studies, giant ellipticals have been the most studied high-redshift galaxies. The first conclusion which can be drawn from the extant work is that at least *some* ellipticals have not evolved significantly during recent times. The combined effects of the color-magnitude relation for ellipticals and of the k-effect guarentee that the very reddest galaxies seen on the sky at any apparent magnitude will be giant ellipticals. Hamilton (1985) has chosen a sample of such galaxies with J magnitudes between 18.0 and 24.0 and has obtained redshifts for 26. Using the amplitude of the 4000 A break as a population indicator, a measure very similar to the rest frame U-B color, he has shown that the populations do not change out to a redshift, $z = 0.8$. His galaxies, at least, have not evolved during that time; but it must be emphasized that, because of the color selection, this is a very special sample which discriminates against evolving objects and his conclusions cannot be extended to the typical giant elliptical.

Somewhat more typical are the brightest members of rich clusters of galaxies (BCM) which have been extensively studied because of their potential use in the determination of q_0. Crane (1975) analyzed the spectra of 28 BCM's with $0.07 < z < 0.46$ obtained by Gunn and Oke (1975). He found evidence for a significant bluing of the spectra with increasing redshift, reaching 0.25 mag at 3700 A by $z = 0.45$. However, Wilkinson and Oke (1978), using the same data supplemented by additional unpublished spectra of fainter galaxies in the same clusters, saw no evidence for evolution; but they did find a large scatter in the SED's of the galaxies at every redshift. Kristian, Sandage, and Westphal (1978) found that the observed B-V colors of a similar sample were consistent, out to $z = 0.8$, with the redshifted SED of M87, again suggesting no evolution.

The more recent survey of Gunn, Hoessel, and Oke (1986), which includes many clusters at redshifts $0.5 < z < 0.8$, has substantially increased the number of BCM's known at high redshift, but the implications for evolution remain controversial. Oke (1983), using 3 different spectral indices sensitive to the presence of young stars, finds that most of the galaxies are similar to nearby ellipticals. A minority are significantly bluer, but Oke claims that the fraction of anomalously blue objects is independent of redshift. Gunn (1986) however, has concluded, from the same data, that the number of peculiar objects does increase with z.

Most powerful, steep-spectrum radio galaxies are also giant ellipticals, only slightly fainter, on average, than the brightest members of rich clusters (Sandage 1972). Like the typical giant elliptical, they tend to be found in dense environments (Prestage 1983), although most are not in rich clusters. Thanks, in large part, to the work of Spinrad and collaborators, radio galaxies have been observed at higher redshifts than have normal BCM's. Recent studies of the color evolution of radio galaxies have been made by Lilly and Longair (1984), Djorgovski, Spinrad, and Marr (1985), and Eisenhardt and

Lebovsky (1986). All three studies show that the range of color increases with redshift. At all redshifts, the majority of objects have colors similar to those expected for a non-evolving giant elliptical. But, at all redshifts, a minority have colors significantly bluer than this, and both the fraction of such objects and their blueward extent, increase with z.

Radio galaxies have often been thought unreliable tools for cosmological studies because of their obvious peculairities. However, most nearby examples have stellar populations indistinguishable from those of normal ellipticals (Sandage 1972) and the recent work of Eisenhardt and Lebovsky (1986), in which a subset of the Gunn et. al. (1986) BCM's were photometered in the same manner as the radio galaxies, suggests that, at every epoch, BCM's and radio galaxies have similar color distributions.

Although not conclusive, I believe that the data favor a picture in which most BCM's are very quiescent objects, evolving only through the aging of their very old stellar populations. A minority, whose numbers have decreased with time, possess younger populations. From broad band colors alone, it is difficult to establish whether these consist of large numbers of intermediate age stars or small numbers of young stars. However, since the evolutionary time scale of any stellar population is of the order of its age, the rapid disappearance of the blue BCM's in recent times favors the latter possibility. We may be witnessing, at earlier epochs, an increasing frequency of bursts of star formation in otherwise normal ellipticals, fueled, perhaps, by infalling gas or gas-rich galaxies. Later, I will suggest a possible mechanism for this process.

CLUSTER GALAXIES

The brightest members of clusters are not typical galaxies. They are not only larger, but, because they usually sit at the bottom of the cluster potential well, they are subject to a number of processes which are unimportant for the average galaxy. Several of these, including cooling flows and the infall of gas-rich galaxies, have the potential for affecting the stellar populations of the central galaxy. The entire population of clusters represent a sample of galaxies which is somewhat more representative. But it is not entirely representative: it has been known for decades that the populations of rich, centrally concentrated clusters are very deficient in spiral and irregular galaxies. This fact may be looked at from two very different perspectives. Butcher and I (Butcher and Oemler 1978) showed that the galaxy population in the cores of rich clusters was well correlated with the degree of central concentration of the cluster. Clusters with no central concentration, whose mean radial profiles resemble that of a uniform density sphere, have almost as many spirals as does the field, but clusters with any significant degree of concentration have very few. On the other hand, Dressler (1980) showed that galaxy populations are a function of the *local* galaxy density, the fraction of spirals decreasing at high density. Implicit in these two views are two very different ideas for the origin of galaxy populations, but the implications for clusters are the same: the cores of rich, centrally concentrated clusters contain no spirals.

This peculiarity can be turned to our advantage. Such clusters are easy to identify even at great distances. One redshift is sufficient to determine the redshifts of all the cluster members. Moreover, we know that at the present epoch the populations of these clusters consist entirely of elliptical and S0 galaxies. Thus, any departure of the colors of the distant cluster galaxies from the colors of a population of E's and S0's will be direct evidence for evolution.

Butcher and Oemler (1984a) assembled photometry of 33 clusters of galaxies with redshifts between 0.003 and 0.54. Of the 33, 29 were centrally concentrated clusters. In each cluster we selected a population within uniform limits of absolute magnitude and distance from the cluster center. After correction for contamination by foreground and background galaxies, we defined the fraction of blue galaxies, f_B, to be the fraction of galaxies with rest frame B-V colors at least 0.2 mag bluer that the mean color of E and S0 galaxies of the same absolute magnitude. (It should be noted that this is a very conservative definition, which excludes many galaxies with stellar populations distinctly different than those of the E/S0's.) At low z, values of f_B range from 0.02 in the typical concentrated cluster, to about 0.1 in some unconcentrated clusters, to 0.4 for field galaxies. At redshifts z > 0.1, f_B in the concentrated clusters begins to rise, reaching a value of 0.25 at z = 0.5. With the exception of one anomalously red cluster, Cl0016 at z = 0.54, there is no evidence for a real variation in f_B among the concentrated clusters at any one redshift, but the uncertainties are large, and some cluster-to-cluster variation in population could be present .

Couch and Newell (1984, Couch 1882) have studied 14 clusters, from the very different perspective of Dressler's morphology-local density relation. Using that relation to predict the distribution of Hubble types in their clusters, and using a relation between Hubble type and color determined from nearby galaxies, they have compared the predicted color distributions to those determined from their photometry. At high redshift, they find a consistent excess in the number of blue galaxies over that predicted. This orthogonal approach leads to the same conclusion to which we came: the number of blue galaxies has decreased by about 25 per cent since the epoch seen at z = 0.5.

A quick review of Fig. 1 will convince the reader that the results of these two studies are not what would be expected in the standard model. Galaxies with the colors of today's E's and S0's should have evolved much less since z = 0.5 than is observed. Rather than accept such precipitous evolution, many have preferred to ascribe the results to some systematic error, and one such possibility comes readily to mind. The contamination of the cluster fields by foreground and background galaxies increases systematically with redshift, and becomes very important by z = 0.5. Since field galaxies are, on average, bluer than cluster galaxies, an underestimate of the field contamination would produce a systematic increase of f_B with z very similar to what is observed.

Fortunately, such an effect is easily testable if one has redshifts of potential cluster members. Seven of the clusters which we included in our sample have now been

TABLE 1

Field Contamination of Blue Cluster Galaxies

Cluster	Estimated Contamination	
	From Photometry	From Spectra
Cl0024+1654	2.5 / 8 = 31%	2 / 10 = 20%
Cl1447+2619	1.7 / 7 = 24%	1 / 6 = 17%
3C295	5.8 / 10 = 58%	4 / 8 = 50%
Abell 223	20 / 27 = 74%	6 / 8 = 75%
Abell 963	22 / 39 = 56%	8 / 17 = 47%
Abell 2111	25 / 50 = 50%	9 / 18 = 50%

studied spectroscopically: Cl0016 (z = 0.54) by Dressler (1986), 3C295 (z = 0.46) by Dressler and Gunn (1983), Cl0024+1654 (z = 0.39) by Dressler, Gunn, and Schneider (1985), Cl1447+2619 (z = 0.37) by Butcher and Oemler (1984b), and Abell 223 (z = 0.21), Abell 963 (z = 0.21), and Abell 2111 (z = 0.23) by Lavery and Henry (1986). The results for the six with significant blue populations are summarized in Table 1, in which I compare the degree of contamination of the blue cluster population by foreground and backgroundgalaxies, as estimated by Butcher and Oemler (1984a), and as determined spectroscopically. The agreement is excellent, eliminating systematic contamination errors as the source of the observed evolution.

The spectra contain other information besides cluster membership. As was first pointed out by Dressler and Gunn (1983), the spectra of the cluster members divide into four classes. The spectra of most of the red galaxies are indistinguishable from those of nearby ellipticals. The spectra of many of the blue galaxies resemble those of nearby spirals, but the majority do not. Among nearby spirals, there is a well-defined relation between the color of a galaxy and the strength of its emission and absorption lines (*q.v.* Dressler and Gunn 1982). This is readily understandable in the standard picture of galaxy evolution, in which the present-epoch color of a galaxy is uniquely related to its entire evolutionary history and therefore, its total stellar population. Compared to this

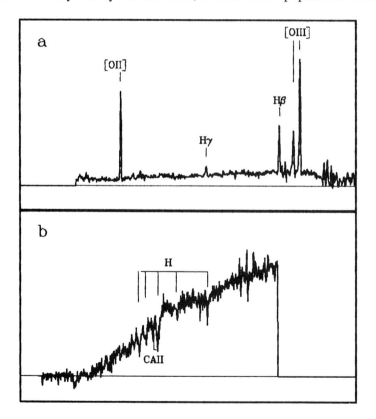

FIGURE 2. a- Spectrum of BO48, a strong emission line galaxy in Cl0024+1654, obtained with the KPNO Cryogenic Camera. b- Spectrum of object b in Cl1447+2619 (Butcher and Oemler 1984b), obtained with the same instrument.

relation, some of the distant blue galaxies have anomalously strong emission lines, similar to nearby Markaryan and starburst galaxies. One example, from the cluster Cl0024+1654, is shown in Fig. 2a. Such objects are uncommon nearby, comprising only a few percent of the galaxy population. Although most of these spectra are probably the result of star formation rather than non-thermal processes, one definite and several probable Seyfert galaxies are known among the distant cluster members, a higher proportion than is seen nearby.

Another class of high redshift blue galaxies has been called by Bothun and Dressler (1986) "post-starburst galaxies." One example, from Cl1447+2619, is shown in Fig. 2b. They have moderately red continua, no emission lines, and an absorption spectrum which combines the features of an early-type galaxy with those of an A star. Unless drastic changes are made to the metallicity or IMF of a galaxy, such a spectrum can only be produced by a large burst of star formation in the recent past, hence the name.

The distribution of these three spectral types among the distant clusters appears significant. The available data are summarized in Table 2 and include those mentioned earlier as well as observations of AC103 by Sharples, Ellis, Couch, and Gray (1985). Although the numbers in each cluster are small, the distribution of the types does not seem to be uniform. For example, the 3C295 cluster contains only post-starburst and high-excitation spectra objects, including one Seyfert I, while Cl0024+1654 contains mostly normal spirals.

TABLE 2
Spectral Properties of Blue Cluster Galaxies

| Cluster | z | f_B | Number of Galaxies | | |
			Normal	HSB	HE
Abell 963	0.21	0.10	3	4	2
Abell 223	0.21	0.19	0	2	0
Abell 2111	0.23	0.16	3	5	1
AC 103	0.31		4	2	0
Cl1447+2619	0.37	0.36	10	1	0
Cl0024+1654	0.39	0.16	11	3	0
3C295	0.45	0.22	0	3	3
Cl0016+16	0.54	0.02	2	9	2

Also noteworthy is that Cl0016, while containing very few blue galaxies (by the Butcher and Oemler 1984a definition) *does* contain a number of post-starburst galaxies, indicating that its population, too, differs from that of nearby clusters. Ellis, Couch, MacClaren, and Koo (1985), using intermediate band photometry, found that these same objects had unusual spectral energy distributions. Many similar objects are found among the population of *red* galaxies in A370 (Ellis 1986), suggesting that in this cluster, and perhaps in others, the population of galaxies with spectral energy distributions different from those of nearby E/S0's may be much larger than has been deduced from broad-band colors.

It is becoming clear that the nature of the blue cluster galaxies cannot be established solely with the kinds of data so far obtained. In particular, we badly need information on the structure of the galaxies, since the spectral energy distributions could be consistent with a wide variety of objects, including almost normal spirals, peculiar ellipticals, and interacting galaxies. Adequate data must await the HST, but a first attempt has been made from Mauna Kea by Thompson (1986). Using a CCD frame of Cl0024+1654 taken in excellent seeing, Thompson was able to deconvolve the radial light distributions of 9 blue galaxies into $r^{1/4}$ law bulges and exponential disks and so obtain bulge-to-disk ratios. Of the nine, seven had intermediate bulge-to-disk ratios similar to those of nearby S0's, one had the small bulge-to-disk ratio typical of a spiral, and one appeared to be an elliptical. In this case, at least, one can conclude that most of the blue galaxies are *not* bursting ellipticals.

FIELD GALAXIES

When we try to understand these observations of galactic evolution, a critical question will be whether the evolution observed in cluster galaxies is representative of what is happening in the general galaxy population, or whether it is a process driven by the cluster environment. The most straightforward way to answer this is by direct observations of the evolution of field galaxies. (By "field galaxies" I shall mean the general galaxy population of the universe, which is dominated by galaxies which are not members of rich clusters).

Such a course is straightforward, but very difficult. Unlike clusters, the redshifts of field galaxies must be obtained one by one. Also, since the field today contains a wide range of galaxy types, we cannot establish the present-epoch properties of any individual high-redshift galaxy. In the overall mix of galaxy types, the evolution of proto-E and S0 galaxies, which we are observing, presumably,in clusters, may well be swamped by the poorly constrained evolution of the larger population of spirals. Despite these difficulties, several redshifts surveys of faint field galaxies have been undertaken. None has yet been completed, but some information is already available. There is some indication that the fraction of "post-starburst" galaxies is not as high in the field as it is in clusters (Koo 1986a, Ellis 1986). On the other hand, it appears that the emission line properties of field galaxies may evolve in a way similar to that observed in clusters. Figure 3 presents some preliminary data from Ellis and Broadhurst (1986) on the distribution of equivalent widths of [OII] 3727, in the AAT sample of nearby field galaxies (Bean *et. al.* 1983), and at high redshifts. There are many more galaxies with strong emission lines in the latter sample, and the range of line strengths is similar to that observed by Dressler, Gunn, and Schneider in the 3C295 and Cl0024+1654 clusters.

Besides the straightforward but very difficult technique of redshift surveys, there exists another approach which can provide information on the evolution of field galaxies. Since the magnitude of a galaxy is an increasing function of redshift, the color-magnitude distribution of a faint sample of galaxies contains information on the stellar populations of high-redshift galaxies. Such data are very much more easily obtained than are redshifts, and a number of attempts (the first being that of Kron 1980) have been made to use photometry of faint galaxies to constrain galactic evolution.

Unfortunately, such data is very much harder to interpret than to obtain. The observed color-magnitude distribution of galaxies is a function of H_0, q_0, and of the

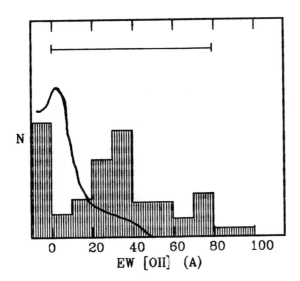

FIGURE 3. The distribution of equivalent widths of the [OII] 3727 line in nearby field galaxies (smooth curve) and in the high-redshift field sample of Ellis and Broadhurst (shaded histogram). The horizontal bracket represents the range in line strength seen in galaxies in the Cl0024+1654 and the 3C295 clusters.

present-day SED, luminosity function and space density of each galaxy type, as well as of galactic evolution. Color is a function of redshift as well as SED, and galaxies at a wide range of distances can have the same apparent magnitude. An apparently red galaxy may be an intrinsically red nearby object, or an intrinsically blue distant one. An increase in the number of blue galaxies at faint magnitudes may be the result of galactic evolution or of a large population of nearby faint blue galaxies.

Despite these problems, most workers seem to agree that the observed color-magnitude distribution of galaxies fainter than 21 magnitude is inconsistent with an unevolving model. How much evolution is required is less clear. Most interpretations of the data have been based on the analytical models of galaxy evolution calculated by Tinsley (1980) and Bruzual (1981, 1983). These models are, at the same time, both too flexible and too restrictive. They are too restrictive in that they assume constant metallicity and IMF, and star formation histories which follow simple analytical forms such as power laws or exponentials. However, even with such limitations to the possible histories of the stellar populations in galaxies, the number of degrees of freedom in these models is much larger than can be effectively constrained by the limited amount of information available in the apparent color-magnitude distribution.

This problem has been partially resolved by recent work of Koo (1986b), who has constructed pairs of color indices from linear combinations of 4 photometric bands. In the resulting two-color planes, the locus of the colors of galaxies of all Hubble types at one redshift is almost perpendicular to the locus of the color of one galaxy at varying redshifts, thus effectively removing the degeneracy of intrinsic color and redshift. Koo finds that his results are not consistent with the standard model of galactic evolution, and that the increase in the fraction of blue galaxies with redshift is similar to that seen by Butcher and Oemler (1984a) in clusters.

DISCUSSION

The problem we face is that of assembling the very incomplete results obtained so far into a consistent picture of the evolution of stellar populations in galaxies. One aspect which may, I think, be readily understood is the evolution of giant ellipticals. It is reasonable to think that the bursts of star formation which were apparently common in BCM's and radio galaxies at earlier epochs could have been produced by infalling gas, in the form of primordial gas clouds or of gas-rich galaxies. With the possible exception of cooling flows onto central galaxies, the members of dense clusters have no such source of gas today, and bursts of star formation are rare. Whatever its explanation, there is strong evidence that clusters at earlier epochs had a much larger population of gas-rich galaxies in their cores, providing a natural source of fuel for the greater star formation activity observed in their central galaxies. In this picture, the observed evolution of BCM's and radio galaxies is a secondary process, an accidental byproduct of the evolution of the gas content of other cluster galaxies.

It is the evolution of those galaxy populations which is the fundamental problem. The standard model cannot account for the rapid changes seen in their properties since the epoch observed at $z = 0.5$. Neither can it explain the peculiar spectra of many of the high-redshift blue galaxies. As was mentioned earlier, the crucial question is whether the observed evolution of cluster galaxies is typical of the broader galaxy population or is a product of the cluster environment. There is considerable evidence that, whatever the mechanism driving the evolution, the most important physical parameter is epoch. Butcher and I found that the spread in the galaxy populations in compact clusters at any one epoch was no larger than expected from observational errors alone, despite a considerable variation in the richness, central concentration, and density of the clusters. What little data are available on unconcentrated clusters suggests a parallel evolution of their populations, as does Koo's (1986) analysis of the field galaxy color-magnitude distribution. If confirmed by more data, these latter two observations would also imply a mechanism which was universal, rather than limited to the cores of rich clusters.

Nevertheless, it is clear that the cluster environment *can* affect the properties of galaxies: the very existence of spiral-poor clusters is a demonstration of that fact. There exists a substantial literature on the anaemic and gas-poor spirals in clusters, and the hot intracluster gas observed in the X-rays provides a natural mechanism, through ram-pressure stripping, for effecting changes in the gas content of cluster galaxies. Also, while the numbers of anomalous galaxies may not vary much from cluster to cluster, Table 2 demonstrates that the *nature* of the peculiarities does, suggesting a cluster-specific process.

A number of such processes have been proposed, many of them based on some variation of ram-pressure stripping. One of the more interesting is ram-pressure induced star formation, suggested by Dressler and Gunn (1982) and Bothun and Dressler (1986). If a spiral galaxy falls into a dense, relaxed cluster, its interstellar medium will be shocked by the ram-pressure produced by the galaxy's motion through the intracluster medium. The final result of this may be to remove the galaxy's gas, but the immediate result may be to precipitate a burst of star formation. In this model, galaxies with strong, high-excitation emission line spectra are assumed to be those currently undergoing a shock-induced burst of star formation, "post-starburst" galaxies are those in which the process has just completed, and the apparently normal spirals are simply milder examples of the same process.

The problem with any cluster-specific mechanism is "why then, but not now?" We see clusters with a wide range of dynamical ages at every redshift. The clusters which have been observed at $z = 0.4$ appear to be identical to those seen nearby in every aspect but galaxy content. What-ever mechanism is proposed to produce blue galaxies at earlier epochs, or eliminate them today, should work equally well at both times. Bothun and Dressler invoke a general evolution of gas content in galaxies, and the fact that clusters collapsing today are of lower density that those which collapsed at $z = 0.4$. However, it is not clear that either of these effects is strong enough the produce the drastic change which is observed in the galaxy content of compact clusters.

Another problem with explaining the peculiar galaxy spectra by a burst of star formation is that of timescales. The lifetime, after a burst, of a strong-emission line spectrum should be a few times 10^7 years, that of a "post-starburst" spectrum, a few times 10^8 years. These times are so short compared to the age of a cluster of galaxies that it is very difficult to understand why almost every cluster which we observe at high redshift contains significant numbers of blue galaxies.

I cannot claim to have a solution to all of these mysteries. However, *if* the evidence for rapid evolution of galaxies outside of rich clusters is confirmed, it is clear that we must look for a process which operates *within* individual galaxies, rather than one dependent on their external environment. Although it has yet to be demonstrated, I think that Larson's bimodal star formation model, described elsewhere in this volume, holds great promise. By decreasing the formation rate of massive stars more rapidly than that of low-mass stars, one may be able to produce galaxies which had the spectral peculiarities and high star formation rates observed in the distant cluster galaxies, but which could have evolved sufficiently rapidly to become the S0's seen in the same clusters today.

REFERENCES

Bean, A.J., Efstathiou, G., Ellis, R.S., Peterson, B.A., and Shanks, T. 1983, *Mon. Not. Roy. Astr. Soc.*, **205**, 605.
Bothun, G.D., and Dressler, A. 1986, *Astrophys. J.*, **300**, 57.
Bruzual, G. 1981, PhD thesis, University of California, Berkeley.
Bruzual, G. 1983, *Astrophys. J.*, **273**, 105.
Butcher, H.R., and Oemler, A. 1978, *Astrophys. J.*, **226**, 559.
Butcher, H.R., and Oemler, A. 1984a, *Astrophys. J.*, **285**, 426.
Butcher, H.R., and Oemler, A. 1984b, *Nature*, **310**, 31.
Couch, W.J. 1982, PhD thesis, Australian National University.
Couch, W.J., and Newell, E.B. 1984, *Astrophys. J. Suppl.*, **56**, 143.
Crane, P. 1975, *Astrophys. J. (Letters)*, **198**, L9.
Djorgovski, S., Spinrad, H., and Marr, J. 1985, in *New Aspects of Galaxy Photometry*, ed. J.L. Nieto (Berlin:Springer-Verlag), p193.
Dressler, A. 1980, *Astrophys. J.*, **236**, 351.
Dressler, A., and Gunn, J.E. 1983, *Astrophys. J.*, **270**, 7.
Dressler, A., Gunn, J.E., and Schneider, D.P. 1985, *Astrophys. J.*, **294**, 70.
Dressler, A. 1986, preprint.
Eisenhardt, P.R.M., and Lebovsky, M.J. 1986, preprint.
Ellis, R.S., Couch, W.J., MacClaren, I., and Koo, D.C. 1985, *Mon. Not. Roy. Astr. Soc.*, **217**, 239.
Ellis, R.S. 1986, private communication.
Ellis R.S., and Broadhurst, T. 1986, private communication.
Gregg, M.D. 1984, Ph.D. thesis, Yale University.
Gunn, J.E., and Oke, J. B. 1975, *Astrophys. J.*, **195**, 255.
Gunn, J.E., Hoessel, J.G., and Oke, J.B. 1986, preprint.
Gunn, J.E. 1986, private communication.
Hamilton, D. 1985, *Astrophys. J.*, **297**, 371.
Koo, D.C. 1986a. private communication.
Koo, D.C. 1986b. preprint.
Kristian, J., Sandage, A., and Westphal, J.A. 1978, *Astrophys. J.*, **221**, 383.
Kron, R.G. 1980, *Astrophys. J. Suppl.*, **43**, 305.
Lavery, R.J., and Henry, J.P. 1986, *Astrophys. J.*, in press.
Lilly, S.J., and Longair, M.S. 1984, *Mon. Not. Roy. Astr. Soc.*, **211**, 833.
Oke, J.B. 1983, in *Clusters and Groups of Galaxies*, eds. F. Mardirossian, G. Giuricin, and M. Mesetti (Dordrecht: Reidel), p 99.
Prestage, R.M. 1983, in *Clusters and Groups of Galaxies*, eds. F. Mardirossian, G. Giuricin, and M. Mesetti (Dordrecht: Reidel), p 559.
Sandage, A.R. 1972, *Astrophys. J.*, **178**, 25.
Searle, L., Sargent, W.L.W., and Bagnuolo, W.G. 1973, *Astrophys. J.*, **179**, 427.
Sharples, R.M., Ellis, R.S., Couch, W.J., and Gray, P.M. 1985, *Mon. Not. R. Astr. Soc.*, **212**, 687.
Tinlsey, B.M. 1980, *Astrophys. J.*, **241**, 41.
Thompson, L. 1986, *Astrophys. J.*, **300**, 639.
Wilkinson, A., and Oke, J.B. 1978, *Astrophys. J.*, **220**, 376.

DISCUSSION

BAUM: I am a bit uncomfortable with a situation in which the present time appears to be "special". The fraction of blue galaxies in clusters was plotted against redshift, rising from roughly f = 0 at z = 0. If we reverse the abscissa and label it "time", the data are telling us that the fraction of blue galaxies in clusters, which was substantial in the past, has been decreasing with time and just happens to be hitting zero today. This seems a somewhat remarkable coincidence, particularly if we note that the approach to f = 0 does not appear to be asymptotic. Nor is f = 0 defined in a wholly arbitrary way; it represents the absence of a blue tail on an otherwise near-normal looking distribution of the colors of galaxies in clusters.

OEMLER: This would be a very serious concern, if the results did imply that our epoch was special, but I don't think that they do. Firstly, since galaxies are steadily using up their gas, they must run out sometime, why not now? And, in fact, f reaches 0 at about z = 0.1, i.e. about 10 percent of the age of the universe ago, not a very close coincidence. Secondly, if we look, not at concentrated clusters, but at unconcentrated clusters or the field, I think we are seeing a parallel evolution, in both of which cases f will reach 0 some time in the distant future.

SCHOMMER: Various investigators have shown that these effects are still occurring today, at least in the richest clusters. Kennicutt, Bothun, and Schommer, and Bothun and Dressler have shown objects in Coma which appear to have enhanced line emission, perhaps due to infall through the intracluster medium. Bothun and Dressler have discussed the orbital parameters required for these galaxies to be infalling for the first time. In Coma, beyond about 1.5 degrees radius, there are many gas-rich galaxies, but the effects are occuring in a much smaller fraction of the total cluster population, with a much longer time-scale now. Thus I don't think we happen to be living at a special time, except after the epoch of densest cluster formation.

WHITMORE: Is there any spatial information for the blue excess frequency? That is, if you break the sample into the 50% nearest the cluster center and the 50% in the outer regions, does this affect the results.

OEMLER: The blue galaxies are invariably much less centrally concentrated than the red ones. And, if there is a central lump of bright galaxies, as in Cl0024+1654, they are always red.

KENNICUTT: Can you comment on the degree to which you can exclude systematic biases in the populations from the selection for distant clusters.

OEMLER: The only potentially serious bias of which I am aware is the fact that a cluster will look richer if it is really the superposition of two clusters along the line of sight. If the clusters are at very different redshifts, the composite cluster will have an artificially wide color distribution. However, the redshift work to which I referred shows that this has not occurred in our clusters.

BURBIDGE: For the radio galaxies with large redshifts which are 3CR sources, it is important to take into account the fact that, since they are powerful non-thermal sources, we can reasonably expect that the colors and energy distributions and optical spectra are determined not only by stellar evolution but also by the effect of non-thermal components: relativistic particles, hot gas, maybe X-ray emission and optical synchrotron radiation. Some of these may be as important as stars in determining the integrated properties.

OEMLER: That could be, but it seems to be true that the radio galaxies, at any redshift, have the same colors as the normal giant ellipticals.

DJORGOVSKI: I disagree with Burbidge's comment about possible contribution of non-thermal light in 3C sources. First, all of their radio power comes from the lobes, with hardly any core contribution. Their images are extended, show no nucleation, and no color gradients. The line emission ([OII] and Lyman alpha) is also extended, not nucleated. Their line emission is typically of low ionization, and the star formation required by their colors and magnitudes is sufficient to power it. The IR colors and magnitudes of 3C sources up to $z = 1$ are indistinguishable from those of brightest cluster members. Thus, we a fairly confident that the light which we are measuring reflects evolution of their stellar populations.

I think that it is an overstatement that the data of Heckman et. al. on low-redshift 3C sources show most of them to be *very* peculiar. I have seen those data, and there are morphological peculiarities, such as shells, or dust lanes, but they are often subtle, rather than dramatic.

KOO: In support of Burbidge's concern that the 3CR radio sample studied by Spinrad and Djorgovski may not be representative of normal gE's in terms of color evolution with redshift, the mJy radio sample published by Kron, Koo, and Windhorst 1985 shows no evidence for bluer colors to redshifts beyond 0.8 (Windhorst, Koo, Spinrad 1986) for identifications likely to be luminous early-type galaxies.

All of the clusters with redshifts beyond 0.3 in Oemler's f_B vs redshift figure were selected from blue plates, with the exception of the "deviant" cluster 0016+16 found by Richard Kron on a deep 4-m red plate. Since 0016+16 was not recognizable as a cluster on equally deep blue plates, such clusters *not* exhibiting an excess of blue galaxies may actually be much more common in samples chosen to be less biased to the blue.

OEMLER: I think that the bias in negligible. In our typical cluster, about one quarter of the galaxies are a few tenths of a magnitude brighter in the blue than the ellipticals. Thus the entire cluster is perhaps 5 percent brighter than it would have otherwise been. On the other hand, I understand that Cl0016+16 was chosen for study specifically *because* it appeared unusually red. Thus it is probably the only biased object in the sample.

LEQUEX: Cluster X-ray emission is an indicator of a hot intracluster medium which may affect galaxy evolution. Is there any correlation between evolution and X-ray emission in your cluster sample.

OEMLER: There is no correlation with X-ray luminosity. There is one with X-ray temperature, in the sense that the hottest clusters, like Cl0016+16, have the reddest galaxies, but the correlation is not statistically significant.

DRESSLER: (In response to a question about subclustering). For the few clusters in which we have about 30 cluster members, the redshift distributions are quite irregular. They do not resemble the Gaussian distribution of a relaxed cluster like Coma. This is a small sample, but it seems to favor the notion that sub-clustering is important. In particular, the post-starburst galaxies usually populate the wings of the distribution. Their high velocities with respect to the cluster mean is one piece of evidence for the ram-pressure induced star formation picture, where a subcluster of gas-rich galaxies collides with the intracluster medium.

In this connection, I want to add that, in addition to the declining gas fractions with time, the fact that clusters collapsing today are less dense, and thus should have a less dense ICM, may explain why these outbursts are not common today.

KING: In line with both of Alan Dressler's remarks, I should like to note that even today only the dense clusters are virialized. As you move out to significant lookback times, it is quite natural that smooth structure should be confined to even denser clusters.

OEMLER: But our high-redshift clusters are structurally identical to our low redshift ones. (Presumably because only relaxed clusters are easy to identify at high redshifts.)

DRESSLER: In the discussion so far we have interpreted the blue galaxies in the distant clusters as likely to be the analogs of blue gas-rich galaxies of today, i.e. late-type spirals. In fact, we have no idea whether these phenomena are associated with earlier Hubble types, even ellipticals. This is obviously an area where HST observations will be decisive.

SPECTRAL EVOLUTION OF GALAXIES: A THEORETICAL VIEWPOINT

Alvio Renzini

Dipartimento di Astronomia, Università di Bologna
and
Osservatorio Astronomico di Bologna

1. INTRODUCTION: WHAT KIND OF ASTROPHYSICS?

Several motivations and strategies coexist in the study of the spectral evolution of galaxies. The most ambitious is perhaps one in which the information contained in the integrated spectra of distant galaxies is decoded to infer galaxy ages. For example, deriving the look back time (LBT) to galaxies and/or galaxy ages for any given redshift. Such an astrophysical chronology, i.e. mapping of redshift vs. time in two directions (LBT vs. z, and age vs. z) is obviously of the greatest cosmological interest.

However, in another motivation one would like to understand other astrophysical processes in galaxies, such as star formation histories, merging, accretion, stripping, and the like. A first important point to realize is that, unfortunately, all these mentioned processes (when involving continuous addition of young stars) ruin galaxies as usable "clocks" for the first cosmological applications. Indeed, bursts of star formation, cannibalism of nearby galaxies, etc. depend, not unlike terrestrial weather conditions, on a number of fortuitous circumstances which make virtually predictionless a theoretical approach. This is to say that only the "simple" case of a passively evolving stellar population can in principle be modelled with some confidence, so as to get theoretical spectral energy distributions (SED) as a function of age (or LBT) which could be used for dating purposes. Therefore, when dealing with the spectral evolution of galaxies a preliminary question naturally comes to mind: are there clock-galaxies? i.e. galaxies in which star formation/addition ceased shortly after the "beginning". Are (some) ellipticals such galaxies? Clearly, the answer to this question determines the kind of astrophysics that we can actually develop.

Beatrice Tinsley, to whom this meeting is dedicated, had such a wide intellectual curiosity so as to certainly embrace all aspects of the spectral evolution of galaxies. But, I believe, when she pioneered this field her first motivation was for the potential cosmological applications. Quite in the same mood, I will now try to review the current state of the research on the spectral evolution of galaxies, with emphasis on the stellar model ingredients, which indeed provide the measure of the time coordinate.

2. GALAXY MODELS AND THE LAST MAJOR EPISODE OF STAR FORMATION IN ELLIPTICALS

What is the typical age of the bulk stellar population in ellipticals? While reviewing attempts at answering this crucial question the opportunity will be taken for a closer look to model stellar populations, their possible drawbacks, and the perspectives at improving their reliability.

It is generally accepted that "most" of the light emitted by elliptical galaxies comes from "old", metal rich main sequence and red giant stars. However, the synthetic spectrum of a simple minded population model, containing just these two ingredients, exhibits three deficits compared to the spectrum of a typical giant elliptical: 1) below ~2000 Å, where the galaxy flux increases for decreasing wavelength, while the model's flux drops precipitously; 2) between ~2000 and ~4500 Å, and 3) above ~1 μ. where in both cases the model flux falls short. A flux deficit implies that the real galaxy contains stellar species which are not included in the simple model: one can then refer to a hot star, a warm star and a cool star deficit, respectively.

Although all three spectral ranges are potentially important for the age problem, it is certainly the ~3300 → ~4500 Å range which is most directly connected to the dating procedures, as turnoff stars dominate. I will therefore restrict the discussion only to the "warm star deficit". In order to do so, it is convenient to refer to a specific set of models, like those of Gunn, Stryker and Tinsley (1981, hereafter GST). Figure 1 shows the GST stellar ingredients, namely: 1) an empirical lower main sequence, 2) the Yale (uncalibrated) isochrones (Ciardullo and Demarque 1977, hereafter CD) transformed (as in GST) according to the prescriptions of Johnson (1966), 3) the "old disc giant" empirical red giant branch (RGB), 4) the upper AGB "addition", that GST introduced to cope with the cool star deficit, and 5) the blue optionals, like the "young", "truncated young", "evolved blue straggler" and horizontal branch (HB) additions, introduced to cope with the warm star deficit.

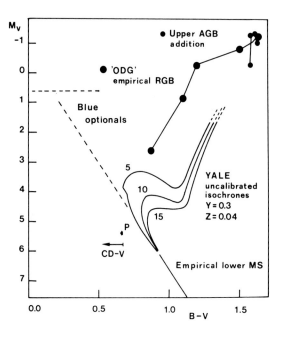

Figure 1. The actual stellar ingredient in the Gunn, Stryker and Tinsley (1981) population synthesis. Isochrones are labelled by age in Gyr. The arrow labelled CD-V gives a typical turnoff color shift between Ciardullo and Demarque (1977) and VandenBergh (1983) isochrones. The arrow labelled P gives the size of the correction applied to CD isochrones by Pickles (1985).

Figure 6 in GST illustrates the case: the 10 Gyr old simple model (no blue optionals) exhibits a very strong warm star deficit, which is greatly reduced when any kind of blue optional is included.   GST conclude that the bulk population in ellipticals is very old ($\gtrsim$ 10 Gyr), with a turnoff color B-V $\simeq$ 0.8, but that a warm stellar component is required.   The nature of the warm population remains open to discussion: either a trace young population, or an old blue straggler (and HB) component.

This conclusion has been questioned by O'Connell (1986a), who noticed the mismatch between CD isochrones and the "old disk giant" RGB apparent in Figure 1 (see also Figure 1 in Tinsley and Gunn 1976).   O'Connell argues that the RGB adopted by GST "is bluer by ~0.3 mag in (B-V) than would be normal for the age and metallicity assigned".   Therefore, when shifting the RGB to the proper, redder location one is forced to pick up a younger isochrone, if one wants to maintain an acceptable fit to the observed SED.   This is basically the argument for favouring young turnoff ages (5-8 Gyr) for gE nuclei.

2.1   Envelope Convection and the Age of Galaxies

The CD isochrones, widely used in a number of applications, were obtained from the grid of evolutionary sequences computed by Mengel et al. (1979), in which a mixing length to pressure scale height of unity was adopted ($\alpha = \ell/Hp$ = 1).   As soon noted by Demarque and McClure (1977), CD isochrones are "far too cool" compared to globular cluster RGB's (by ~0.2 mag in B-V), and main sequences.   They report that the choice α = 1.6 gives a reasonable fit to observed RGB locations (see also Buonanno, Corsi and Fusi Pecci 1981, VandenBerg 1983).

It is worth emphasizing that this operation of determining the best value of α is not only legitimate, but absolutely indispensable before venturing into applications of stellar models to the spectral evolution of galaxies. Indeed, this <u>calibration</u> of the models is to some extent analogous to the flat-fielding procedures used in CCD photometry, and is seemingly essential if accurate results are to be achieved!

In the frame of the mixing length theory, the mixing length ℓ is a free parameter to be determined <u>a posteriori</u>, and for <u>convenience</u> it is <u>assumed</u> proportional to Hp.   There is therefore no a priori constraint on the value of α, and there is no guarantee for one particular value of α to be good in all situations.   Variations of α are physically equivalent to variations in the efficiency of convection in carrying out heat, and therefore they affect the radius (effective temperature) of the models, but have little effect on the structure of the deep interior (and then on the luminosity).   Changes in model temperature (color) induced by changes in α are extremely sensitive to model characteristics.   For instance, for dwarfs hotter than ~7000 K the envelope convection is either very shallow or non-existent.   Correspondingly, model radius and temperature (color) are insensitive to α (the envelopes are practically radiative).   At the opposite extreme, i.e. dwarfs cooler than ~4000 K, thanks to high subphotospheric densities convection is very efficient (~ adiabatic), and again the structures are almost insensitive to variations of α.   On the other hand, at intermediate temperatures (around 5000-6000 K) dwarf models have a superadiabatic envelope of considerable thickness, and correspondingly exhibit a maximum sensitivity to α.   Finally, the superadiabatic convective envelope of red giants encompasses most of the stellar volume, and correspondingly the radius of RGB models is very sensitive to the adopted treatment of convection. Some of these trends can be clearly appreciated by looking at Figure 3 in VandenBerg (1983), which shows how the temperature shift is far from being constant. To further illustrate the case, Figure 1 shows the shift in B-V at turnoff between CD and VandenBerg (1983) isochrones (for t = 12 Gyr, Y = 0.3, Z = 0.01). Most of this shift is due to the variation of α from 1 to 1.6. Note that Δ(B-V) ≈ -0.15 is equivalent to an increase in age from 5 to 15 Gyr (!).

In conclusion, the temperature calibration of the models is a very complex affair, which by no means can be reduced to the application of a constant temperature shift. Moreover, this calibration is methodologically equivalent to <u>synchronizing</u> the clock used to infer ages from spectra, and to set the <u>rate</u> at which this clock runs (see below). This calibration must therefore be regarded as one of the crucial, central ingredients in a theoretical approach to the spectral evolution of galaxies.

While most researchers in this field are aware of the existence of the calibration problem, it appears to me that its importance has been generally overlooked. For example, GST, O'Connell (1980) and Bruzual (1983) used straight CD isochrones without any correction. Others, e.g. Mould (1978), Aaronson et al. (1978), Frogel, Persson and Cohen (1980), Pickles (1985) and Arimoto and Yoshii (1986a, b), have applied corrections which are too small and/or constant, and therefore cannot be regarded as generally satisfactory.

Figure 2 compares VandenBerg (1983) and O'Connell (1980, Table 3) turnoff color vs. age, the latter being derived from uncalibrated CD isochrones. Note that the two calibrations are in good agreement insofar as envelope convection is unimportant, but start to diverge for turnoff colors redder than B-V≈0.45,

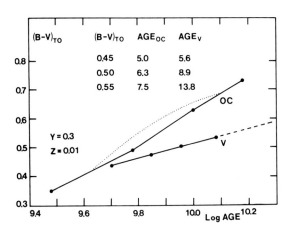

Figure 2: The turnoff color versus log Age for the O'Connell (1980) and VandenBergh (1983) calibrations. The dotted line gives the turnoff color as obtained from CD isochrones using Johnson (1960) color-temperature calibration.

reaching more than a factor of 2 age discrepancy at $(B-V)_{TO}$ = 0.6. Note in particular that the two slopes are different: this is equivalent to say that the two clocks do not run with the same speed! The SED of a model galaxy constructed using CD isochrones evolves much faster than the one constructed with VandenBerg isochrones.

Finally, concerning gE nuclei, the use of VandenBerg (α = 1.6) isochrones of the proper metallicity would probably give an age of ~15 Gyr (rather than ~5-8 Gyr), even when using a RGB ~0.3 mag redder than the one adopted by GST.

2.2 The Case of M32

It is not clear whether M32 is a typical elliptical galaxy without other specifications, or just a typical dwarf elliptical companion of a giant spiral. In any case, it has been extensively studied, thanks to its prox- imity. In particular, O'Connell (1980) synthetic study concludes that the turnoff stars belong to spectral type F8 and the metallicity is nearly solar. When using uncalibrated CD isochrones this corresponds to an age of ~5 Gyr, about 1/3 of the age currently assigned to galactic globular clusters. However, the age of the globulars is currently derived using methods other than that of synthetic spectra, and the suspicion arises that the discrepant ages might be an artifact of the use of different dating methods. The neces- sity of calibrating synthetic methods using template populations has been illustrated by Renzini and Buzzoni (1986), and those arguments will not be repeated here.

Although it would be preferable to apply to galactic globulars precisely the same procedure used by O'Connell on M32, I will try here a somewhat

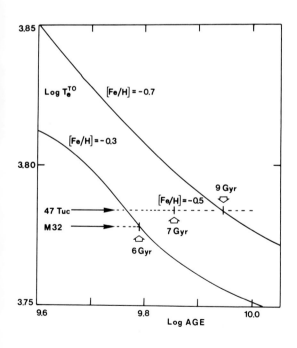

Figure 3: The turnoff effec- tive temperature versus log Age for uncalibrated CD isochrones. From them nearly the same age is inferred for M32 and 47 Tuc (see text).

different experiment. From Table II in Johnson (1966) one derives B-V = 0.54 and T_e = 6000 K for F8 dwarfs, i.e. for the turnoff stars in M32. For their metallicity I adopt [Fe/H] = -0.3 (Z = 0.01). This is less than used by O'Connell, but hardly inconsistent with the available evidence. For instance, Rose (1985) estimates [Fe/H] = -0.25 and Mould (1978) gives [Mg/H] = -0.3. One can now enter with these values into Figure 3, and from uncalibrated CD isochrones one derives an age of ~6 Gyr. One can now repeat the same procedure for the metal rich globular 47 Tuc: from the excellent CCD color-magnitude diagram of Harris and Hesser (1985) one gets the turnoff color $(B-V)^{TO}$ = 0.56, or $(B-V)^{TO}_o$ = 0.52 after reddening correction. Inserted in Figure 3, again using Johnson's Table II, this gives an age of ~7 Gyr adopting [Fe/H] = -0.5, and ~9 Gyr adopting [Fe/H] = -0.7. The conclusion is that turnoff colors indicate nearly the same age for M32 and 47 Tuc, rather than a factor of 3 difference. Concerning the absolute value, the young derived age is a consequence of the use of uncalibrated CD isochrones: from Figure 2 one sees that with $(B-V)^{TO}$ = 0.54 and Z = 0.01 VandenBerg isochrones give an age of ~14 Gyr.

In conclusion, the case of M32 illustrates the well known difficulty in deriving absolute ages from colors (spectra) which are extremely sensitive to metallicity (cf. caveats in O'Connell 1980, 1986a,b, Iben and Renzini 1984, Renzini 1984, Renzini and Buzzoni 1986). It also illustrates that, due to the flatter run of $(B-V)^{TO}$ vs. age in VandenBerg isochrones, small differences in the adopted turnoff colors give large age differences.

2.3   The Case of Giant Ellipticals

The canonical metallicity assigned to gE galaxies is near twice solar, or Z = 0.04. For this metallicity and an age of 15 Gyr CD isochrones give a turnoff temperature log T_e = 3.7095 which translates to $(B-V)^{TO}$ = 0.86 using Johnson's Table II. One is then in a range of temperatures where turnoff color is most sensitive to the adopted value of mixing length. A choice α = 1.6, which gives a good fit to globular cluster RGB's, would most probably reduce the 15 Gyr isochrone turnoff color to less than 0.70, the value currently indicated by empirical synthesis of gE galaxies (cf. O'Connell 1986a, Pickles 1985). I conclude that there is no compelling evidence for intermediate age ($\lesssim$ 8 Gyr) stars in gE galaxies, and that a proper choice of the mixing length goes some way towards eliminating the apparent elliptical/globular cluster age discrepancy, although the calculation of high-Z isochrones (for various α values) would be appropriate to check more precisely this point. Moreover, the proper inclusion in empirical synthesis of minority and trace stellar ingredients could only reinforce this conclusion. Some related aspects are discussed next.

2.4   The Dispersion of Stellar Metallicities

In the previous section I have followed the same procedure used in the cited synthetic studies: i.e. identify a turnoff color, adopt a typical metallicity and then enter into isochrone tables to sort out an age. I believe, however, that this procedure may be badly misleading, as real galaxies certainly exhibit a range of stellar metallicities, and in population synthesis it is never legitimate to replace a distribution with its average. For example, let us consider an old population (#1) with a certain metallicity distribution $\phi(Z)$, and average <Z>, and a second (#2) population which is homogeneous in metallicity, with Z = <Z>. The integrated spectra of the two populations will be markedly different, the larger the metallicity dispersion in population #1. Indeed, the warmest main sequence and turnoff stars in #1 are <u>warmer</u> than their counterparts in #2 and, seemingly the coolest RGB stars in #1 are <u>cooler</u> than their counterparts in #2. Therefore, population #2 will lack a group of warm and a group of cool stars compared to #1, and its integrated spectral energy distribution will be correspondingly <u>narrower</u>, thus developing a warm and cool star deficit syndrome (cf. beginning of Section 2).

It is worth emphasizing that the expected dispersion in metallicity is substantial. For example, the gE model galaxies of Arimoto and Yoshii (1986a, b) exhibit a full dispersion of about 2-3 orders of magnitude, similar to that of the spheroidal component of spirals (cf. Whitford and Rich 1983, Mould, this meeting, and Mould and Kristian 1986, for a recent observational result on M31). The effects on the SED of model populations of such large metallicity dispersions remain to be quantitatively investigated. Qualitatively, one can anticipate some expected trend:

1)   a single metallicity synthetic fit to a real galaxy underestimates its age (the synthesis code will ask young stars to mimic metal poor ones). Turnoff ages derived in this way should then be regarded only as lower limits.

2)   When spectral features are compared, there is no guarantee that population #1 and population #2 will have the same set of indices. For instance, suppose Mg_2 (1) = Mg_2 (2), there is no guarantee that also the Hβ, Sr II, CN indices and the like will be identical for the two populations. Actually, we may expect that the larger the metallicity dispersion in #1, the larger the difference in all other indices.

In conclusion, there is much to be investigated about synthetic models of realistic populations, and the question of the intrinsic metallicty dispersion is certainly one of those deserving top priority. I believe that at this stage we have more to learn from home made numerical experiments (simulations), than from empirical synthetic optimizations of specific galaxies based on dramatically incomplete libraries.

2.5   Chemical Complicacies

In two recent extensive studies (Burstein et al. 1984, Rose 1985) a variety of spectral indices are constructed from the integrated spectra of galactic globulars, M32, and M31 globulars. Burstein et al note that M31 globulars, M32 and ellipticals follow a different Hβ vs. Mg_2 correlation compared to galactic globulars. They also note that M31 globulars exhibit stronger CN indices compared to either galactic globulars or ellipticals at the same Mg_2, and favour younger ages for the studied extragalactic systems as a possible interpretation. Rose notes that M32 has weaker Sr II lines compared to metal rich galactic globulars, and from a comparison with individual solar neighborhood dwarfs and giants he infers that M32 is more "dwarf-dominated", and then younger than galactic globulars.

The conclusions of these spectroscopic studies and the synthesis results of O'Connell (1980) and Pickles (1985) lend mutually support for the notion of a younger age (5-8 Gyr) for the bulk population of elliptical galaxies.

However, my impression is that no alternatives to younger ages have been really explored in the attempt to account for the observed differences. On the other hand, when dealing with spectral features of particular elements their abundance is certainly at least as relevant as the age, and when dealing with stellar systems other than the solar neighborhood the default assumption of solar elemental proportions ($[X_i/Fe] = 0$) becomes highly suspicious.

These considerations give the opportunity for a glance towards a virtually unexplored field: the detailed chemical evolution of spheroidal components (bulge-halo) and elliptical galaxies. In this connection, it is worth referring to the qualitative concept of "chemical trajectories" for any given pair of elemental abundances X_i and X_j, as the actual path in the (X_i, X_j)-plane followed by a given (portion of) galaxy in the course of its evolution. For example, the Oxygen vs. Iron abundance in the galactic halo, Strontium vs. Magnesium in M32, etc.

Chemical trajectories in the (X_i, X_j)-plane depend on (at least) three relevant timescales: the lifetime of the stars producing the elements i and j (τ_i and τ_j, respectively), and the characteristic timescale of star formation τ_G, defined as the time taken to complete the bulk of star formation. Clearly, only if both τ_i and τ_j are $\ll \tau_G$ the instantaneous recycling approximation is valid, and the X_i^j/X_j ratio is independent of τ_G. When at least one of the stellar lifetimes becomes comparable to τ_G then the chemical trajectory becomes sensitive to the actual star formation history. An example of this situation can be found in the study of Matteucci and Greggio (1986) about the relative role of type I and II supernovae on the variation of the O/Fe ratio in the galactic halo. According to a widespread prejudice that I like, elliptical galaxies and spheroidal components are characterized by τ_G values in the range between ~0.1 to ~1 Gyr. We should then expect quite a variety of chemical trajectories, depending on the actual τ_G of each object, even in the ideal case of a universal initial mass function, and universal τ_i values. This is to say that during early galactic evolution chemistry and dynamics are inextricably interwoven, and that elemental proportions in stars are a product of the detailed history of star formation. Alien proportions are too be expected in alien worlds, while solar proportions should be regarded only as the product of the particular "chemical trajectory" followed by the solar neighborhood.

In this framework, the weardness of M32 and M31 cluster family (relative to the galactic family) could be attributed to different chemical trajectories, rather than to younger ages. For instance, the HB-Mg$_2$ anomaly could result from a systematically different horizontal branch morphology vs. metallicity, particularly in metal rich clusters (average HB B-V around 0.3, rather than 0.7; cf. Section 4c)i) in Burstein et al 1984). In turn, such a systematic difference could be produced by a systematically different run of [O/Fe] vs. [Mg/H], i.e. a different chemical trajectory. Space Telescope observations of M31 clusters will provide adequate color-magnitude diagrams, and will then check their possible HB anomaly.

Concerning other indices, it is worth noting that both C, N and the s-process elements (like Strontium) are probably produced in relatively long lived intermediate mass stars, and that C and N are subject to further processing in low mass stars (cf. Truran, this volume). Their use as age indicators should then be discouraged, owing to the complex, and partly still unknown, physical processes involved in their production, reprocessing and distribution (cf. Danziger, Matteucci and Kjär 1985).

In conclusion, my impression is that convincing evidence has not yet emerged for M31 globulars, M32 and gE nuclei being significantly "younger" than galactic globulars. This leaves the possibility open for a cosmological use of E galaxies.

3. CONCLUSIONS (SO MUCH WORK TO DO ...)

The construction of theoretical models for the spectral evolution of stellar populations is straightforward in principle, and painfully messy in practice. This is due to the large variety of ingredients which are required to construct usable models, and to the shortage on the market of several essential ingredients. Model makers are then forced to use e.g. incomplete libraries of stellar spectra and uncalibrated isochrones, and to omit unfamiliar evolutionary stages. Some of these aspects have been discussed in Renzini and Buzzoni (1986), and here I will limit myself to a few comments on the problem of the isochrones.

As already mentioned in Section 2.1, the steeper the $(B-V)^{TO}$ vs. age relation, the faster the spectral evolution of a model population. Furthermore, the slope of the $(B-V)^{TO}$-age relation depends on the adopted mixing length, parameter α, in such a way that the smaller α, the larger the slope. The most widely used set of spectral evolutionary models (Bruzual 1983) is based on $\alpha = 1$ CD isochrones, and therefore these models evolve too fast compared to properly calibrated ($\alpha \simeq 1.6$) models. In particular, in color-redshift diagrams they depart too promptly from purely redshifted (unevolving) SED's.

The problem of the evolution of stellar populations and evolutionary population synthesis clearly tends to orient a number of otherwise disparate stellar studies. Moreover, the perspective of the potential cosmological applications provides additional scientific interest and motivation to this sector of the astronomical research. At this stage, having recognized some of the limitations of the early generation of models, the necessity arises for planning the next generation of evolutionary population syntheses. The requirements it should fulfil are manyfold; actually so numerous and diverse to perhaps require a virtually "industrial" approach. I will list here some among the most urgent of them:

1) Calibration of theoretical stellar evolutionary sequences using template stellar populations. The calibration should involve at least the 3 main free parameters: a) mixing length for the outer convection, b) mass loss during the RGB and AGB phases, and c) overshooting beyond formal convective boundaries. The calibration should cover the largest possible fraction of the age-metallicity plane, with the caveat in mind of possible higher composition dimensions, and consider all evolutionary stages (MS, RGB, HB, AGB, and POST-AGB).

2) Compilation of empirical and/or model spectral libraries of reasonable completeness, for all relevant wavelengths, temperatures, gravities, and compositions.

3) An assessment of the role played by interacting binaries (blue stragglers and merged binaries, white dwarf accretors, etc.) on the integrated spectrum of populations, in particular in the Blue-UV and near-IR.

4) Calibration of population synthesis codes using the same template populations involved in testing the stellar evolutionary sequences (cf. Renzini and Buzzoni 1986).

By templates populations one intends here resolvable star clusters of sufficient richness, like globular clusters in the Galaxy, the Magellanic Clouds, and the other local group galaxies, implemented by the galactic bulge of the Milky Way and M31 for the crucial high metallicity case. In order to exemplify this calibration approach I would like to mention two studies which are more familiar to me. The first is the photometry of a complete sample of 10000 stars in M3 (Buonanno et al. 1986a), which has for the first time allowed the accurate determination of the relative contributions of the stars in the various evolutionary stages to the integrated cluster light. I hope that this archetipal study will soon be followed by many others.

A second study involves Buonanno, Corsi, Fusi Pecci, Greggio, Sweigart and myself, and concerns the so called RGB phase transition in 0.5-1.5 Gyr old populations, i.e. the development of an extended RGB when stars with a degenerate helium core first appear in a population. As part of this project ~100 stellar evolutionary sequences have been constructed, and good CCD color-magnitude diagrams have been obtained for over a dozen globular clusters in the Magellanic Clouds (for a preliminary account see Buonanno et al 1986b). The choice of these particular clusters and age range is deliberately based on the consideration that elliptical galaxies at redshift ~1 (if they exist!) may be dominated by stars in the corresponding mass range. Our preliminary conclusion is that the RGB phase transition does not cause a large jump in the integrated B-V, while it probably does so in the integrated B-R or V-K. Unfortunately, this tends to weaken the possible detection and cosmological use of the RGB phase transition in high redshift elliptical galaxies (cf. Renzini and Buzzoni 1986). However, we are currently enjoying a new kind of "experimental photometric synthesis" (EPS) made possible by the CCD data. Cluster integrated magnitude and colors can easily be obtained from such data, and particular stars or groups of stars can be artificially inserted in, or eliminated from the frames. This procedure allows a direct, experimental assess of the relative contribution of stars in the various evolutionary stages to the cluster integrated light and colors. A fruitful future for EPS of globular clusters is anticipated.

REFERENCES

Aaronson, M., Cohen, J. G., Mould, J. R., Malkan, M. 1978, Ap. J., **223**, 824.
Arimoto, N., Yoshii, Y. 1986a, Spectral Evolution of Galaxies, ed. C. Chiosi and A. Renzini (Dordrecht: Reidel), p. 309.
Arimoto, N., Yoshii, Y. 1986b, preprint.
Bruzual, A. G. 1983, Ap. J., **273**, 105.
Buonanno, R., Corsi, C. E., Fusi Pecci, F. 1981, M.N.R.A.S., **196**, 435.
Buonanno, R., Buzzoni, A., Corsi, C. E., Fusi Pecci, F., Sandage, A. 1986a (in preparation).

Buonanno, R., Corsi, C. E., Fusi Pecci, F., Greggio, L., Renzini, A.,
    Sweigart, A. V. 1986b, Mem. S. A. It., (in press).
Burstein, D., Faber, S. M., Gaskell, C. M., Krumm, N. 1984, Ap. J., **287**, 586.
Ciardullo, R. B., Demarque, P. 1977, Trans. Astr. Obs. Yale Univ., **33**, 1.
Danziger, I. J., Matteucci, F., Kjär, K. 1985, Production and Distribution of
    CNO Elements (Garching: ESO).
Demarque, P., McClure, R. D. 1977, The Evolution of Galaxies and Stellar
    Population, ed. B. M. Tinsley and R. B. Larson (New Haven: Yale Univ.
    Obs.), p. 139.
Frogel, J. A., Persson, S. E., Cohen, J. G. 1980, Ap. J., **240**, 785.
Gunn, J. E., Stryker, L. L., Tinsley, B. M. 1981, Ap. J., **249**, 48.
Harris, W. E., Hesser, J. E. 1985, Dynamics of Star Clusters, ed. J. Goodman
    and P. Hut (Dordrecht: Reidel), p. 81.
Iben, I. Jr., Renzini, A. 1984, Physics Reports, **105**, 329.
Johnson, H. L. 1966, Ann. Rev. Astron. Ap., **4**, 193.
Matteucci, F., Greggio, L. 1986, Astron. Astrophys., **154**, 279.
Mengel, J. G., Sweigart, A. V., Demarque, P., Gross, P. G. 1979, Ap. J.
    Suppl., **40**, 733.
Mould, J. R. 1978, Ap. J., **220**, 434.
Mould, J. R., Kristian, J. 1986, Ap. J., **305**, 591.
O'Connell, R. W. 1980, Ap. J., **236**, 430.
O'Connell, R. W. 1986a, Spectral Evolution of Galaxies, ed. C. Chiosi and A.
    Renzini (Dordrecht: Reidel), p. 321.
O'Connell, R. W. 1986b, Publ. Astron. Soc. Pacific, **98**, 163.
Pickles, A. J. 1985, Ap. J., **296**, 340.
Renzini, A. 1984, Observational Tests of Stellar Evolution Theory, ed. A.
    Maeder and A. Renzini (Dordrecht: Reidel), p. 21.
Renzini, A., Buzzoni, A. 1986, Spectral Evolution of Galaxies, ed. C. Chiosi
    and A. Renzini (Dordrecht: Reidel), p. 135.
Rose, J. A. 1985, A. J., **90**, 1927.
Tinsley, B. M., Gunn, J. E. 1976, Ap. J., **203**, 52.
VandenBerg, D. A. 1983, Ap. J. Suppl., **51**, 29.
Whitford, A. E., Rich, R. M. 1983, Ap. J., **274**, 723.

DISCUSSION

O'CONNELL: Alvio, we seem to have a problem, or at least a misunderstanding. Three comments...

1) The age calibration I showed this morning <u>did</u> include correction for the mixing length.   The shift was a constant with temperature, but that wouldn't create large errors near the Sun, which is the calibration point.

2) I believe Pickles (who obtained main sequence turnoff colors for gE's like mine) used fully corrected Yale isochrones.

3) When I checked VandenBerg newer isochrone calculation, after the color correction he found necessary, I estimated that $(B-V)^{TO} \simeq 0.50$ for 5 Gyr. A few $0^m_.01$ shift would not significantly affect the situation.   I do agree that small color shift for $(B-V)^{TO} \gtrsim 0.65$ will significantly effect age estimates, but for warmer turnoffs near 0.50, effects of temperature corrections on age estimates should be smaller.

RENZINI: Yes, I think we have a problem, in particular with gE's.   Concerning your comments:

1) I was referring to your 1980 paper, in which the problem of the mixing length is discussed, but CD isochrones are used at face value, i.e. without any correction.

2) Pickles has used Twarog (1980, <u>Ap. J.</u>, **242**, 242) corrections, which are too small for main sequence and turnoff stars at the relevant temperatures, i.e. $\Delta\log T_e \simeq 0.005$ rather than ~0.03 as in VandenBerg vs. CD isochrones at 15 Gyr (see also Figure 1).   The use of better calibrated isochrones will then give considerably older ages than those inferred by Pickles for the studied galaxies.

3) I agree.

O'CONNELL: I now realize what accounts for at least part of our disagreement.   Your ages for M32 assumed Z = 0.01 and $(B-V)^{TO}$ = 0.54.   Mine assumed Z = 0.02 and $(B-V)^{TO}$ = 0.50.   The observations are more consistent with the latter value in my view.

RENZINI: Maybe, but the point is that such small differences in the assumptions produce large differences in the derived age.

BURSTEIN: a)   I fully agree with your desiderata for the future.   Our group and others are engaged in programs to begin such analyses.   b)   However, I have trouble with your interpretation of M32.   The IUE spectrum from 2400 to 3200 Å is very close to F8V.   The IUE spectral atlas shows that this region of spectrum behaves quite non-linearly from F5 to G5, so that this F8V sp. type is reasonably well determined.   F8V, near solar metallicity = 5 Gyr?

RENZINI: Yes, if F8V is $(B-V)^{TO}$ = 0.50 and "near solar" is solar.   No, if F8V is $(B-V)^{TO}$ = 0.54 and "near solar" is half solar, in which case the age is ~14 Gyr.   Besides this, I think that a dispersion in metallicity will favour older ages.

BURSTEIN: We did consider the effect of a mixture of metallicities in the M31 globulars.   If metal poor stars produce the Balmer line excess, they would require the CN strength in the more metal rich stars to be even bigger than the improbable values already implied.   This did, and does, seem improbable to me.

RENZINI: Before yesterday my guess was that the Balmer line excess was perhaps due to your M31 clusters being on average much brighter than galactic ones, i.e. as bright as ω Cen, and as ω Cen could have a large metallicity dispersion. However, Leonard Searle explained to me that in his sample the Hβ anomaly persists even in fainter clusters. That is why I favour a "second parameter" effect on the HB morphology. Concerning CN, I don't see the connection between CN strength and age, and in any case I would not use CN indices as age indicators before understanding the origin of CN anomalies in galactic globular cluster stars.

KING: I am concerned about the idea of applying the Whitford-Rich abundance spread to cases like M31 and M32. Although their results are intrinsically very interesting, they refer to a region that is 600 pc away from the plane, in a bulge that is smaller than that of M31. There may be bad contamination.

RICH: I did some calculations on the percentage of what you would expect for a halo and disk contribution, based on a rotation curve. The contribution turns out to be relatively small, really on the order of 10%, if there is a central component of the galaxy, as required to support the rotation curve.

KING: Based upon the Bahcall-Soneira model, which does not support the galactic rotation curve and has not been tested in toward the galactic center, one expects 50% disk contamination. However, if we use the Bahcall, Schmidt, and Soneira (1983) central component, the total contamination (disk + bulge) falls to 10%.

RENZINI: I see. I referred to Whitford-Rich results in support to the notion of a large metallicity dispersion in ellipticals.
EDMUNDS: To answer Dr. King's point the simple "closed box" model gives (by definition!) the simplest first guess at the evolution of a system, and it automatically gives a wide spread in abundances in the stars. It is only by invoking other processes--such as inflow (as in the solar neighborhood), outflow or change in IMF that the metal abundance distribution in the stars can be made narrower.

GUNN: What is the full range in turnoff color for the closed box model?

RENZINI: With Arimoto and Yoshii metallicity spread for a gE model [Fe/H] ranges from ~-1.5 to +0.9, and $(B-V)^{TO}$ would correspondingly range from ~0.40 to over 0.70.

DRESSLER: You mentioned that if Bruzual had used calibrated isochrones he would have predicted less dramatic evolution. Does this, in your opinion, greatly strengthen the case for significant evolution in high redshift galaxies?

RENZINI: I think so.

AARONSON: I think some of the large variations in V-K of cloud clusters that you referred to are simply due to stochastic variations in small numbers of carbon stars. This point may also be relevant to the M31 clusters, as it's my impression that their IR colors do not show evidence of C star light, and therefore they are probably much older than ~5 Gyr.

RENZINI: I fully agree with both your points.   Concerning the V-K of Magellanic clusters I remember Jay Frogel arguing that a jump in this color remain
even when taking AGB stars out.

ROCCA-VOLMERANGE: A calibration of the Yale isochrones on globular clusters
of the Magellanic Clouds can only fit a normal or highly deficient metallicity
stellar population.   It is impossible with such a calibration to fit elliptical galaxies, specially in the red colors.   You should need to calibrate on
a super-metal-rich giant branch!   The "uncalibrated" Yale isochrones correspond to very cold effective temperatures of the red giant branch.   It is for
this reason that models based on such isochrones can fit elliptical
galaxies.   In any case, MC globular clusters cannot be template of a metal-
rich population.

RENZINI: I fully agree.   I feel that we shall need a great deal of
theoretical imagination at the high metallicity end.

GALAXY POPULATIONS: STRUCTURE AND KINEMATICS

K.C. Freeman
Mount Stromlo and Siding Spring Observatories
The Australian National University

INTRODUCTION

This talk will be mainly about our Galaxy, in the context of what we know about other disk galaxies. I will discuss the old thin disk, the thick disk and bulge, the metal-weak halo and the globular cluster system. The young population and the dark corona will not be discussed here.

THE OLD DISK

The old disk is a flat rapidly rotating component which contains most of the galactic luminous mass. The luminosity density of the old disk can be represented by $L(R,z) \propto \exp(-R/h_R - z/h_z)$. The vertical lengthscale h_z comes from star counts, and is in the range 300 to 350 pc (see for example Pritchet 1983). The radial lengthscale is probably between 3.5 and 5.5 kpc. The shorter value comes from comparison of the calculated surface brightness near the sun with the central surface brightness of the disk as observed for other spirals, and also from the radial distribution of HI and CO (see de Vaucouleurs and Pence, 1978). The longer value comes from van der Kruit's (1986) analysis of the Pioneer 10 photometry: his best model gives $h_r/h_z \approx 17 \pm 3$.

The old disk is supported vertically by its velocity dispersion, and radially mainly by its rotation. Near the sun, its velocity dispersion components are $\sigma_z \approx 20$ km s^{-1}, $\sigma_R \approx 40$ km s^{-1}, $\sigma_\varphi \approx 25$ km s^{-1} (see for example Hartkopf and Yoss, 1982; Lewis 1986). The circular velocity is about 220 km s^{-1}, and the asymmetric drift for the old disk in the solar neighborhood is about 15 km s^{-1}. We see that the old disk is relatively cold near the sun. Is this so everywhere ? This is worth knowing, because it bears on the dynamical history and stability of the disk.

Cold disks are unstable to axisymmetric modes if the parameter $Q = \sigma_R \kappa/(3.36G\Sigma) < 1$ (Toomre, 1964), where κ is the local epicyclic frequency and Σ is the surface density, and they may be unstable to bar modes if $Q > 1$. Near the sun, the value of Q is about 1.3. The bar modes can be suppressed by:

(1) having a significant fraction of spherically distributed matter in the inner regions of the galaxy. The dark coronas which give flat rotation curves are probably dynamically important in the outer parts only

(Carignan and Freeman, 1985; van Albada et al, 1985).

  (2) making the disks <u>hot</u> in the inner parts. For example, a K giant disk with $Q \gtrsim 2.5$ in the inner parts (Athanassoula and Sellwood 1986). This option is not obviously attractive because the galactic disk is cold near the sun.

  However, van der Kruit and Searle's (1982) photometry of several edge-on spirals showed that the vertical scale height of their disks are approximately constant with radius. We would then expect the vertical velocity dispersion $\sigma_z \propto \exp(-R/2h_R)$. This has actually been observed for old disks of a few face-on spirals, such as NGC 5247 (van der Kruit and Freeman, 1986). Then, <u>if</u> the ratio σ_R/σ_z is approximately constant with R, the radial velocity dispersion would also follow $\sigma_R \propto \exp(-R/2h_R)$. In the inner parts of the Galaxy, σ_R would then rise to about 120 km s^{-1}, similar to the velocity dispersion of the bulge and the halo.

  The radial variation of the σ_R component is relatively difficult to measure in other disk galaxies, because the measured line of sight velocity dispersion is the projection of at least two of the components $(\sigma_R, \sigma_\varphi, \sigma_z)$ and the variation of the rotational velocity, integrated along the line of sight through the disk. However, in our Galaxy, $\sigma_R(R)$ can be measured directly from radial velocities of individual disk stars that lie in galactic windows.

  Lewis (1986) has observed about 600 K-giants in 7 galactic windows where the reddening $E(B-V) < 0.5$ is already known from distant objects (clusters or galaxies). The windows have galactic latitudes $|b| < 4°$ and galactic longitudes between 186° and 25°. The radial velocities were measured with the AAT fiber system and with the KPNO 2.1-m telescope. The distances of individual stars are estimated from photometric parallaxes.

  The K-giants in Baade's window and the anticenter window show that σ_R varies smoothly between $2 < R < 16$ kpc, and follows closely the dependence $\sigma_R(R) = 122 \exp[-R(\text{kpc})/2h_R]$ km s^{-1}, where $h_R = 4.0$ kpc (preliminary value). This is close to what we would expect for a disk of uniform vertical scaleheight and uniform ratio of σ_R/σ_z, as discussed above. The high central value of σ_R may raise some suspicions that the sample of disk stars in the inner parts of the Galaxy may be contaminated by bulge stars. However, from the Bahcall-Soneira export model, modified to include the known reddening laws in these windows, it can be shown that the observed values of σ_R are significantly affected by contamination of the sample by bulge stars.

  The radial behaviour of Toomre's Q can be estimated from this run of $\sigma_R(R)$, assuming that the surface density is exponential and calculating $\kappa(R)$ from the galactic circular velocity curve. In the inner parts of the disk ($R < 3$ kpc), Q rises above 2.5; from comparison with the theoretical results of Athanassoula and Sellwood, it seems likely that the disk is hot enough in its inner regions to be stable against barlike modes. It is important now to understand the heating processes that led to this high central velocity dispersion and the exponential run of σ_R.

A comment on the asymmetric drift of an exponential disk with an exponential run of σ_R: for such a disk, the calculated asymmetric drift is everywhere less than about 20 km s^{-1}, even near the center where $\sigma_R \approx 120$ km s^{-1}. Lewis's observations show directly that the asymmetric drift is indeed small, at least in the region R > 3 kpc, where the mean azimuthal motion of the old disk stars is well determined. In contrast, an isothermal isotropic population with density distribution $\propto r^{-3.5}$ in a potential with a circular velocity of 220 km s^{-1} is in equilibrium with zero mean rotation if its $\sigma_R \approx 120$ km s^{-1}.

Another comment, on the disklike nature of a disk with a central velocity dispersion comparable to that of the bulge: if the vertical scale height of the old galactic disk is constant with radius, then σ_z rises from about 20 km s^{-1} in the solar neighborhood to about 60 km s^{-1} near the center (assuming a radial scale length of about 4 kpc). In the same interval, σ_R rises from about 40 to 120 km s^{-1}. The disk can remain disklike towards its center, because its velocity dispersion remains anisotropic, with $\sigma_R/\sigma_z \approx 2$.

THE THICK DISK

van der Kruit and Searle (1982) derived vertical surface brightness profiles for several edge on spirals. For galaxies with negligible bulges, the profiles showed only the thin disk component. However, for galaxies like our galaxy, with small bulges, there is also a thick disk component of longer vertical scale height. From star counts at the SGP, Gilmore and Reid (1983) showed that our galaxy appears to have a similar thick disk component. They interpreted their star counts as the sum of two components: a thin disk with scaleheight $h_z \approx 300$ pc and a thick disk of scaleheight ≈ 1500 pc. The local density of this thick disk component is about 2% of the thin disk's.

An important paper by Hartkopf and Yoss (1982:HY) gives insight into the vertical (z) kinematics of the thick disk. They studied the kinematics and abundances of G and K giants at the galactic poles, at distances of up to about 5 kpc. Within a few hundred parsecs of the galactic plane, almost all the stars have abundances in the range 0 > [Fe/H] > -0.5, and their σ_z = 22 ± 1 km s^{-1}; these stars belong to the old thin disk, whose scale height is about 300 pc. At greater heights, there are many metal weak ([Fe/H] < -1) stars which belong to the galactic halo (see below). However there are also many relatively metal rich giants (0 > [Fe/H] > -1) far from the galactic plane, which could not belong to a thin disk population whose density distribution is exponential with a scale height of about 300 pc. We identify them tentatively with the thick disk. For example, in the region 2 < z < 5 kpc, the HY sample has about 30 stars with abundances between 0 and -1. If the Gilmore-Reid normalisation is correct, then only one or two of these stars could belong to the thin disk. The vertical velocity dispersion of these 30 stars is 39 ± 6 km s^{-1}.

This σ_z of about 40 km s^{-1} appears to be characteristic of the thick disk at the solar distance from the galactic center. It is consistent with a population whose scale height is about 1500 pc, and is found for several samples of stars whose abundances lie between the characteristic range for the thin disk (0 to -0.5), and the low values associated with ha stars ([Fe/H] < -1). For example, Rose (1985) identified a sample of nearl luminous G-stars as RHB stars with typical abundances of about -0.7. The σ_z is about 40 km s^{-1}. Another example comes from nearby stars in HY's sam (z < 1 kpc) with abundances between -0.5 and -1; their σ_z = 43 ± 7 km s^{-1}.

If this is all correct so far, then the thick disk appears to k kinematically intermediate, in its vertical dynamics, between the rapidly rotating thin disk ($\sigma_z \approx 20$ km s^{-1}) and the slowly rotating metal weak hald ($\sigma_z \approx 75$ km s-1: see below). The next question is the rotation of the thic disk: is it also rotationally an intermediate population ?

Ratnatunga and Freeman (preprint) studied the abundances and kinematics of K giants in a field at l = 270°, b = 39°. This field is use for estimating the mean rotation of the various populations. Our sample includes stars out to distances of about 25 kpc and with abundances in the range 0 to -2. In particular, it includes metal rich stars, between 2 and kpc above the galactic plane, that must belong to the same population as t HY thick disk stars at the galactic poles. Our velocity data shows a fairl abrupt discontinuity in the kinematical properties of stars at [Fe/H] ≈ -0 The more metal weak stars belong to a non-rotating (halo) population with line of sight velocity dispersion of about 125 km s^{-1}. The more metal rich stars (which in our sample are found in the region 2 < z < 5 kpc) belong t rapidly rotating population with a mean velocity (relative to the LSR) of only 36 ± 14 km s^{-1} and a line of sight dispersion of 50 ± 10 km s^{-1}.

From these observations, we can estimate the radial component c of the velocity dispersion for this thick disk population, and also its asymmetric drift. Some assumptions are needed: (i) our metal rich stars ha the same σ_z as the corresponding HY stars with 2 < z < 5 kpc, (ii) their c $(2)^{1/2}\sigma_\varphi$, (iii) the galactic rotation curve is flat between the solar radiu and their galactocentric distance of 9.3 kpc. It then follows that our thi disk stars have a σ_R = 69 ± 18 km s^{-1} and that their asymmetric drift is on 29 ± 19 km s^{-1} (these numbers are dynamically compatible if the thick disk has a radial scale length of 4 to 5 kpc). The thick disk is kinematically hotter than the thin disk, but both are rapidly rotating populations. The Table below compares present estimates of some parameters for the thin and thick disks.

Disk	h_z (pc)	σ_z	σ_R	asymm drift (km s^{-1})
Thin	350	20	40	15
Thick	1500	45	70	30

I have argued here that the population of metal rich stars with > 2 kpc is rotating almost as rapidly as the thin disk, and I have associa this population with the Gilmore-Reid (GR) thick disk because (i) these st are too far from the galactic plane to be part of the thin disk, if the GR normalisation is correct, and (ii) their σ_z is consistent with the GR scaleheight of 1500 pc. However, the relationship between the thin and thick disks is not yet clear: are they discrete components, or is the thick disk just the higher energy, slightly metal-weaker tail of the disk population ? This question is not just semantic. Its answer is important for understanding how the thick disk formed. For example, did it form during the galactic collapse, when the galaxy was already close to centrifugal equilibrium but had not completely settled in the vertical direction ? Or was it produced after the thin disk had begun to form, perhaps by some secular heating process like spiral arm heating, or by a more rapid process such as heating by a transient bar or by the accretion of satellites (see Quinn and Goodman 1986). More extensive data on the vertical structure of the thin and thick disks, as determined from a homogeneous sample of stars, would be helpful.

More precise information on the shape of the velocity ellipsoid for the thick disk population could be particularly useful in choosing between the possible formation stories, and in understanding why thick disks and bulges appear to go together. Although the table above suggests that the σ_R/σ_z ratios for the thin and thick disks are similar (≈ 2), the uncertainty is still so large that this ratio for the thick disk could be near unity. Consider a thick disk in which $\sigma_R/\sigma_z \approx 1$ and in which the velocity dispersion rises towards the galactic center, reaching a central value similar to that observed by Lewis for the inner parts of the thin disk (about 120 km s^{-1}). The central regions of this thick disk would no longer be disklike, but would probably have the appearance of a small bulge. Rowley has made some preliminary models of such a thick disk population in the potential of a dominant thin disk: these models are able to produce quantitatively most of the known structural and kinematical properties of the thick disk and small bulge of systems like our galaxy.

THE METAL WEAK HALO

The metal weak stars ([Fe/H] < -1) form a spheroidal slowly rotating population. Its surface density distribution, as defined mainly by the globular clusters, is close to an $R^{1/4}$ law (de Vaucouleurs and Pence, 1978); the effective radius of this distribution is about 2.7 kpc. The globular cluster distribution of M31 also follows approximately an $R^{1/4}$ law (de Vaucouleurs and Buta, 1978). The metal weak halo, as defined for example by the globular clusters and RR Lyrae stars, shows no clear radial abundance gradient, and its volume density distribution is fairly well represented by an $R^{-3.5}$ law (Zinn 1985; Saha 1985). Near the sun, the density of this component relative to the disk is about 0.2 percent (Bahcall et al, 1983).

Near the sun, the kinematics of the metal weak halo can be studied from samples of metal weak stars found kinematically (ie from high proper motion catalogs) or non-kinematically (from spectroscopic and photometric surveys and from variability for RR Lyrae stars). To derive the mean rotational velocity motion v_{rot} and velocity dispersion σ from kinematically selected samples, correction for kinematic bias is needed. For example,

Bahcall and Casertano (1986) applied Monte Carlo simulations to Eggen's catalog, and derived v_{rot} = 66 ± 23 and $(\sigma_R, \sigma_\varphi, \sigma_z)$ = 140 ± 12, 100 ± 12, 7 6 km s^{-1}.

Estimates for non-kinematically selected samples are more straightforward. Norris et al (1985) discussed a sample of about 70 spectroscopically selected stars with [Fe/H] < -1.2 and with good UVW velocities, and found v_{rot} = 27 ± 22, and $(\sigma_R, \sigma_\varphi, \sigma_z)$ = 125 ± 11, 96 ± 9, 8 7 km s^{-1}. A more extended sample of non-kinematically selected stars with [Fe/H] < -1.2, which includes RR Lyrae stars, BHB stars, and metal weak giants and dwarfs (Norris, 1986) gives v_{rot} = 37 ± 10, and $(\sigma_R, \sigma_\varphi, \sigma_z)$ = 13 6, 106 ± 6, 85 ± 4 km s^{-1}.

The metal weak halo shows no obvious [Fe/H] gradient. We shoul also ask whether the kinematical properties of this population, near the depend on [Fe/H] ? This is worth knowing, because different pictures of galaxy formation make different predictions about how the kinematics and abundance are related. Norris's metal weak sample ([Fe/H] < -1.2) shows n significant change of v_{rot} and sigma with [Fe/H]. This conclusion appears differ from that of Sandage (see his paper in this volume), who studied t kinematics of a large new sample of nearby kinematically selected halo st The reason for the apparently different conclusions is not yet fully understood. I think it may have to do with the different ways in which th two authors define the halo.

Sandage defines the halo spatially; ie he identifies the halo stars in his proper motion sample as those whose orbits take them above s minimum distance (a few kpc) from the galactic plane. With this definitio the halo sample will include some stars that we would otherwise associate with the thick disk. We recall that the thick disk is fairly cold and rap rotating, and contains stars with abundances in the range 0 to about -1. also know that the metal weak stars ([Fe/H] < -1) belong to a hot, slowly rotating population. Therefore, if one takes subsamples of stars of diffe abundance from a spatially defined halo sample, then the kinematical properties of the subsamples will indeed change with abundance. This spat definition of the halo is entirely valid if one regards the thick disk as last stage of star formation in the halo, after the galaxy had more or le settled radially but had not yet dissipated to a disk.

Norris defines the halo chemically. He argues that there is a fairly abrupt transition in the kinematic properties of stars, at an abundance of about -1. The more metal rich stars are rapidly rotating, wh the more metal weak stars are rotating slowly. The kinematics of globular clusters show a similar behaviour (Zinn, 1985). Norris identifies the hal stars as those whose [Fe/H] values are less than say -1.2. These stars ap to belong to a kinematically homogeneous slowly rotating population. With this definition, most of the thick disk stars are excluded from the halo sample.

Which one of these definitions is more valid ? It probably depends on the reader's preconceptions about galaxy formation. My own

preference at this stage is for the chemical definition, because I am
impressed by the idea that (i) the metal weak slowly rotating halo is the
debris of low mass satellites that were accreted early in the life of the
galaxy (see below), and (ii) the rapidly rotating thick disk comes from
heating of the early thin disk by this accretion process (see Quinn and
Goodman, 1986) and the remarks by Wyse at the end of this paper). If this
view is correct, then the halo and thick disk are dynamically quite distinct,
and should be regarded as separate components; the thin disk and the thick
disk are much more closely related.

      We return now to the sample of nearby spectroscopically selected
stars studied by Norris et al (1985). This sample includes a number of stars
with low eccentricity (e < 0.4) orbits and low abundances ([Fe/H] < -1.2;
such stars were missing from the Eggen et al (1962) compilation. The detailed
kinematics of these stars are interesting. Their mean rotation is high, as
one would expect from their low eccentricities: their v_{rot} = 183 ± 8 km s^{-1}
(compared with v_{rot} = 27 ± 22 km s^{-1} for the whole sample with [Fe/H] <
-1.2). However, their σ_z value is only 44 ± 10 km s^{-1} (compared with 88 ± 7
km s^{-1} for the whole metal weak sample). To which population do these stars
belong ? Their σ_z value of 44 km s^{-1} is characteristic of the thick disk.
Therefore these low-e, low abundance stars could be a metal weak tail of the
thick disk: the presence of such stars in the thick disk would be a useful
pointer to its formation history. Alternatively, they may just be the
kinematical tail of a non-rotating population with v_{rot} ≈ 0 and σ_φ ≈ 100 km
s^{-1}. The low-e stars of this population would be about 2-σ, in the rotational
component of the velocity, from the mean of this population, so stars with
low-e orbits would have predominantly low z-velocities, particularly if the
population has an energy cutoff (like a King model).

      Now we turn to the kinematics of halo objects far from the sun;
data is available now for metal weak giants, RR Lyrae stars, BHB stars and
globular clusters out to large galactocentric distances. The first point is
that the observed velocity dispersion of these objects is remarkably
isothermal (at about 120 km s^{-1}) from near the center out to at least 35 kpc
(Freeman, 1985; Norris, 1986). The next point concerns the shape and
orientation of the velocity ellipsoid for halo stars. Near the sun, we know
that the velocity dispersion components are $(\sigma_R, \sigma_\varphi, \sigma_z)$ ≈ 140, 100, 75 km s^{-1};
the velocity ellipsoid is highly anisotropic and is elongated in the radial
direction. We would like to know the orientation of the velocity ellipsoid
far from the galactic plane: for example, does it point towards the galactic
center or towards the axis of galactic rotation. There is enough data
available now on the kinematics of halo stars, at different distances from
the sun and in different directions, to test some simple kinematic models of
the halo.

      Ratnatunga and Freeman (1986) have tested two such kinematic
models, using our own data and the data of Beers et al (1986) for metal weak
giants, and the data of Pier (1983) and Sommer-Larsen and Christiansen
(1986a) for BHB stars. The models that we tested have the velocity ellipsoid
constant, first in spherical coordinates {ie $(\sigma_R, \sigma_\varphi, \sigma_\theta)$ = 140, 100, 75 km s^{-1}
everywhere} and then in cylindrical coordinates {ie $(\sigma_R, \sigma_\varphi, \sigma_z)$ = 140, 100, 75

km s^{-1} everywhere}. At the SGP, the observed velocity dispersion appears to be approximately constant with distance, with $\sigma_z \approx 75$ km s^{-1} from near the galactic plane up to z ≈ 25 kpc. This is what we would expect for a system that is isothermal and cylindrical in the above sense. Away from the galactic poles, the velocity dispersion data is again well fit by a system which is isothermal in cylindrical coordinates; the model which is isothermal in spherical coordinates is a poor fit to the data.

We have also used the velocity data for distant metal weak giants in our field at l = 270°, b = 39°, to see if the kinematical properties of the outer halo stars with [Fe/H] < -1 change with abundance. The stars between 10 and 25 kpc from the sun were split into two groups: -1 > [Fe/H] -1.4, and [Fe/H] < -1.4. The mean line of sight velocity and the line of sight velocity dispersion for the two samples are very similar: $<v_{los}> \approx 170$ and $\sigma_{los} \approx 125$ km s^{-1}.

In summary, the metal weak halo is a slowly rotating, roughly isothermal system. Its velocity dispersion is anisotropic, with the long axis pointing at the axis of galactic rotation, rather than at the center of the galaxy. For stars with [Fe/H] $\lesssim$ -1, there does not seem to be much change the kinematical properties with abundance.

THE GLOBULAR CLUSTER SYSTEM

Zinn (1985) showed that the galactic globular clusters appear to fall into two groups, defined by their distribution, kinematics and chemical properties. A histogram of the abundance distribution for the clusters shows a bimodal distribution; the two modes separate at [Fe/H] $\approx$ -1. The clusters in the more metal rich mode lie in a rotating disk, with v_{rot} = 152 ± 29 and σ = 71 ± 12 km s^{-1}. Although the scaleheight of this disk is not well determined, the kinematical parameters are within the errors of those for thick disk, as discussed above, and it is tempting at this stage to identify the metal rich clusters with the thick disk. The metal poor clusters belong to a slowly rotating spheroidal population, with v_{rot} = 50 ± 23 and σ = 114 9 km s^{-1}: these values are fairly close to those for the metal weak stellar halo.

It is interesting to compare the shape of the velocity ellipsoid for the metal weak halo clusters and for the metal weak stars. For the stars we know that the velocity ellipsoid is highly anisotropic. Frenk and White (1980) made a kinematic solution for the motions of the clusters, and inferred that the velocity dispersion for the cluster system is close to isotropic. This is an interesting result. If the clusters and the metal weak halo stars are kinematically different, then it suggests that the clusters and the stellar halo have had a different dynamical history, or maybe that the clusters in the most elongated orbits have by now been destroyed. However, Norris (1986) has made a new solution for a larger sample of 73 metal weak clusters ([Fe/H] < -1.2). Although the three components of the velocity ellipsoid are approximately equal, as found by Frenk and White, t formal errors of the solution are large and it is not yet possible to say whether the clusters are indeed kinematically different from the halo stars.

Rodgers and Paltoglou (1984: RP) looked in more detail at the rotation of the globular cluster system. They noted that globular clusters are known to form in the disks of small galaxies like the LMC. In the disk of the LMC itself, globular cluster formation has continued up to the present time, and even the oldest globular clusters in the LMC lie in its disk (Freeman et al, 1983). RP considered the possibility that the globular clusters which are seen now in the galactic halo formed in small disklike satellite galaxies, perhaps like the LMC and SMC, which were then accreted by the galaxy (see also Searle and Zinn, 1978). Although the satellite itself would be tidally disrupted, the globular clusters would probably survive, because they are more tightly bound. In this picture, clusters that came from a common parent may be in a restricted abundance range, and may still have common kinematics. Rodgers and Paltoglou therefore made rotation solutions for clusters in several abundance intervals. The v_{rot} values for most of these abundance intervals were between 40 and 100 km s^{-1} (except for the metal rich clusters, which are of course rotating more rapidly). However, the rotation for the 30 clusters with $-1.3 > [Fe/H] > -1.7$ is retrograde, with $v_{rot} \approx -70$ km s^{-1}. This result is marginally significant.

The idea, that the metal weak globular clusters (and therefore probably the metal weak field stars also; see Silk and Norman 1979) formed in satellite galaxies which were then accreted long ago by our galaxy, is interesting and deserves further work. Quinn and Goodman (preprint) have studied the dynamics of the accretion of satellites by a disk galaxy. They showed that satellites in orbits inclined at less than about 60° to the galactic plane are pulled down towards the plane by dynamical friction, before being tidally disrupted. The vertical energy of their orbits goes into heating of the galactic disk. Satellites in more highly inclined orbits just lose orbital energy until they are disrupted, without being pulled down into the plane. This process could lead naturally to the apparently sharp structural and kinematic discontinuity between the slowly rotating metal weak halo and the rapidly rotating thin and thick disks. It is also possible that the thick disk itself resulted from the heating of the galactic (thin) disk that occurred during the accretion phase.

There is evidence for a few moving groups of metal weak stars in the galactic halo. These are presumably the debris of broken up globular clusters or accreted satellite galaxies. In the solar neighborhood, the Kapteyn's star group is the best known (Eggen, 1979): however, its reality remains contentious. This group appears to have $[Fe/H] \approx -1.8$ and has low orbital angular momentum, so its orbit carries it close to the galactic center. This is at least consistent with the idea that the group comes from some small tidally disrupted system. Recently Sommer-Larsen and Christiansen (1986b) discovered a group of 6 BHB stars in a small region of space, about 4 kpc from the sun. The radial velocities of these stars appear to correlate tightly with their distance. With some simple assumptions about the dynamical nature of halo groups, it is then possible to estimate the orbital parameters for this group. Again, its angular momentum is low, and its orbit takes it within 1 kpc of the galactic center.

REFERENCES

Athanassoula, E & Sellwood, J. (1986). Preprint.
Bahcall, J.N., Schmidt, M. & Soneira, R. (1983). The galactic spheroid.
          Astrophys.J., 265, 730-747.
Bahcall, J.N. & Casertano, S. (1986). Preprint.
Beers, T.C., Preston, G.W. & Shectman, S. (1985). Preprint.
Carignan, C. & Freeman, K.C. (1985). Basic parameters of dark halos in late
          type spirals. Astrophys.J., 294, 494-501.
de Vaucouleurs, G. & Buta, R. (1978). On the distribution of globular
          clusters in the galaxy and in Messier 31. Astron.J., 83, 1383-9
de Vaucouleurs, G. & Pence, W. (1978). An outsider's view of the galaxy.
          Astron.J., 83, 1163-73.
Eggen, O.J., Lynden-Bell, D. & Sandage, A. (1962). Evidence from the motion
          of old stars that the galaxy collapsed. Astrophys.J., 136,
          748-66.
Eggen, O.J. (1979). Intermediate-band photometry of late type stars.
          VIII.Astrophys.J., 229, 158-74.
Freeman, K.C. (1985). The old population. In The Milky Way Galaxy, ed H. va
          Woerden et al, Dordrecht: Reidel.
Freeman, K.C., Illingworth, G.D. & Oemler, A. (1983). The kinematics of
          globular clusters in the LMC. Astrophys.J., 272, 488-508.
Frenk, C.S. & White, S.D.M. (1980). The kinematics and dynamics of the
          galactic globular cluster system. Mon.Not.R.astr.Soc., 193,
          295-311.
Gilmore, G. & Reid, N. (1983). New light on faint stars, III.
          Mon.Not.R.astr.Soc., 202, 1025-1047.
Hartkopf, W. & Yoss, K. (1982). A kinematic and abundance survey at the
          galactic poles. Astron.J., 87, 1679-1709.
Lewis, J.R. (1986). ANU thesis, in preparation.
Norris, J.E., Bessell, M.S. & Pickles, A.J. (1985). Population studies I: t
          Bidelman-MacConnell "metal weak" stars. Astrophys.J.Suppl., 58,
          463-92.
Norris, J.E. (1986). Preprint.
Pritchet, C. (1983). Application of star count data to studies of galactic
          structure. Astron.J., 88, 1476-88.
Quinn, P.J. & Goodman, J. (1986). Preprint.
Pier, J.R. (1983). AB stars in the southern galactic halo, II.
          Astrophys.J.Suppl., 53, 791-813.
Ratnatunga, K. & Freeman, K.C. (1986). Preprint.
Rodgers, A.W. & Paltoglou, G. (1984). Kinematics of galactic globular
          clusters. Astrophys.J., 283, L5-L7.
Rose, J. (1985). Red horizontal branch stars in the galactic disk. Astron.J
          90, 787-802.
Saha, A. (1985). RR Lyrae stars and the distant galactic halo. Astrophys.J.
          289, 310-19.
Searle, L. & Zinn, R. (1978). Compositions of halo clusters and the formati
          of the galactic halo. Astrophys.J., 225, 357-79.

Silk, J. & Norman, C. (1979). Gas rich dwarfs and accretion phenomena in
        early type galaxies. Astrophys.J., $\underline{234}$, 86-99.
Sommer-Larsen, J. & Christiansen, P. (1986a). Radial velocities of blue
        horizontal branch field stars in the outer galactic halo.
        Mon.Not.R.astr.Soc., $\underline{219}$, 537-46.
Sommer-Larsen, J. & Christiansen, P. (1986b). Preprint.
Toomre, A. (1964). On the gravitational stability of a disk of stars.
        Astrophys.J., $\underline{139}$, 1217-38.
van Albada, T.S., Bahcall, J.N., Begeman, K. & Sancisi, R. (1985).
        Distribution of dark matter in the spiral galaxy NGC 3198.
        Astrophys.J., $\underline{295}$, 305-13.
van der Kruit, P. & Searle, L. (1982). Surface photometry of edge-on spiral
        galaxies III. Astron.Astrophys., $\underline{110}$, 61-78.
van der Kruit, P. (1986). Surface photometry of edge-on spiral galaxies, V.
        Astron.Astrophys., $\underline{157}$, 230-44.
van der Kruit, P. & Freeman, K.C. (1986). Stellar kinematics and the
        stability of disks in spiral galaxies. Astrophys.J., $\underline{303}$, 556-72.
Zinn, R. (1985). The globular cluster system of the galaxy, IV. Astrophys.J.,
        $\underline{293}$, 424-44.

DISCUSSION

Rich   How faint are your stars in the direction of the galactic center, ar
do you believe the Bahcall-Soneira model with its two components, disk and
spheroid, rather than including the central component required to support
rotation curve.

Freeman   Our faintest stars are about V = 16. The problem of bulge
contamination that you are alluding to is serious only in the innermost pa
of the galaxy. I am not sure how well the rotation curve is known in this
region, so we prefer to use the BS model directly.

King   Two students of mine, Kate Brooks and George Chui, did theses analy:
densities and motions in high latitudes. I told them to fit with a disk p.
a halo. We were always puzzled that they had to use a halo that was flatte
about 3:1. It now seems that the reason was that their "halo" stars had a
large admixture of thick disk. Now a different point: no-one seems to ag:
when I suggest that the central bulge of our Galaxy is simply what a disk
does when its central regions are dominated by random motions rather than
rotation. But now you show a σ_R in the disk that rises in the center to ju
the value that it needs there. Why then do you still call the bulge a
separate component ?

Freeman   It seems clear that σ_R for the bulge and the inner disk are
comparable. We only know σ_R: I think that the shape of the velocity ellips
will determine whether the distribution is a disk or a bulge near the cent
If the anisotropy of σ near the center is similar to the anisotropy of the
disk near the sun, then probably the distribution will be disk-like.

Janes   There is another observation which may bear on the thick disk probl
There exists a small number of open clusters at distances of about a
kiloparsec from the galactic plane. These clusters are not particularly ol
(the youngest is about 1 Gyr) and they have metallicities roughly in the
range of the thick disk sample. It isn't at all clear how they got there.
These clusters are primarily in the outer disk but with one possible
exception they are not distant enough to be associated with the warp seen
the galactic HI layer. Furthermore the young open clusters define a very f
disk, which does however thicken as one looks outward along the disk.

Pier   Recent measurements of RR Lyrae field stars towards the NGP show no
decrease in velocity dispersion from a few kpc to over 20 kpc - it remains
constant at about 110 km s^{-1}.

Whitmore   The central stellar velocity dispersion of a normal face-on Sc
galay is about 60-80 km s^{-1}. In your modelling of the Milky Way, you requi:
$\sigma \approx 120$ km s^{-1} to stabilise the disk. Are these at odds with each other ?

Freeman   No: it is the radial component of σ that needs to be about 120 km
s^{-1}. The velocity ellipsoid is probably highly anisotropic.

Larson   I'm a litle bothered by your suggestion that the thin disk forms
first, and then the thick disk immediately forms by some process that puffs
up the thin disk. A thin disk that survives to the present time can't begin
to exist until any processes that keep disrupting or puffing up the thin disk
have died out. How is this situation different in practice from a more
continuous formation picture in which the halo forms first, then the thick
disk, and finally the thin disk by a progressive shrinkage of the gas
distribution.

Freeman   I was suggesting the opposite process. The thin disk could be the
first well-defined structure to form. The halo then forms by the accretion of
small metal weak satellites. In the process, the stars that had already
formed in the thin disk at that time are heated (through the dynamical
friction) to become the thick disk that we now see. This accretion phase
probably lasts only a few dynamical times (as Quinn and Goodman
demonstrated). Then the thin disk can continue its evolution undisturbed.

Da Costa   If the thin disk is formed first, what mechanisms are there to
accelerate Zinn's metal rich globular clusters into the thick disk.

Freeman   Satellite accretion, heating by a central bar,and heating by spiral
structure are all possibilities.

Wyse   Bernard Jones and I worked on a model of disk galaxy formation, which
attempted specifically to understand the van der Kruit and Searle and
Burstein observations of thick disks in external galaxies (Astron.Astrophys.,
120, 165, 1983). We investigated the collapse of gas within a nonspherical
slowly rotating dark potential well. In our model, the formation sequence of
luminous components of disk galaxies is that the extreme spheroid forms first
during the rapid collapse along the shortest axis of the potential. The thick
disk forms during the "pancaking" stage, being puffed up by subsequent
relaxation as the radial collapse is completed. The metal rich, rotating
bulge is the central region of the thick disk, while the thin disk forms once
the potential is steady. The Galactic thick disk seems a reasonable fit to
the predictions of our model.

Djorgovski   Is it possible that the distinction between the thin and thick
disks is artificial, namely that they are just two ends of a same population?
In other words, are the dynamical and chemical properties changing smoothly
from one to the other, or are there any discrete jumps ?

Freeman   This is an important question. I don't think we know the answer yet.

Gilmore   An important model-independent result follows from the apparent
anisotropy of the velocity ellipsoid of the metal poor extreme subdwarfs and
giants. If the velocity ellipsoid really is as anisotropic as it appears,
the spheroid isodensity contours must be highly flattened. An axial ratio of
at least 1:3 is unavoidable, even if the total galactic potential is round.
Thus, anisotropy of the velocity ellipsoid is direct evidence that the metal
poor extreme spheroid has a significantly flatter distribution than that
inferred from the globular cluster population.

CONCLUDING REMARKS

James Lequeux
Observatoire de Marseille

We will all agree that the present meeting has been an excellent and very exciting one, due to the uniformly high standards of the review papers, the lively discussions and the wealth of new results which have been presented. Progress has already been quite significant since the Erice workshop one year ago (Spectral Evolution of Galaxies, ed. C. Chiosi and A. Renzini, Reidel, Dordrecht, 1986). It is clear that the proceedings of the present conference will be a most appropriate update of those of the Yale conference (The Evolution of Galaxies and Stellar Populations, ed. B. M. Tinsley and R. B. Larson, Yale University Observatory, 1977) which have been the bible for many of us for almost ten years.

It is an almost impossible task to summarize in a few sentences the rich contents of this meeting, and I will only recall a few highlights--reflecting unavoidably my personal interests, some trends in this very active field, and I will also review a selection of points where the Hubble Space Telescope is likely to carry very significant progress.

We all know that galaxies differ from each other! However some considerable differences between rather similar systems have come out unexpectedly. Leonard Searle, Bob Zinn and Jerry Mould have shown that while no big differences are noticeable in the bulges of the Andromeda Nebula and our Galaxy, their haloes are enormously different. Globular clusters and field stars in the M31 halo have medium metallicities, with a large spread and no clear radial dependence, while those in the halo of our Galaxy have significantly lower metallicities; however the inner globulars in our Galaxy form a clearly different, rotating system, flattened and more metal rich; no such dichotomy seems to exist in M31. The latter type of globulars may have something to do with the thick disk of Gilmore and Reid. The existence of such a disk has been confirmed by the very extensive kinematical study of subdwarfs reported by Alan Sandage: according to him, the thick disk population has intermediate metallicity (0.3 to 1 x solar, with a real spread), a vertical velocity dispersion of about 45 km s^{-1} and lags in rotation behind the thin disk. Ken Freeman has also discussed this component in his brilliant review of galactic kinematics during the last day, and believes that the differences in opinion about the separate existence of the thick disk may just reflect differences in opinion about its (still unsettled) origin: slow dissipative collapse or acceleration of stars through their encounters with interstellar clouds. I have been also particularly interested in Freeman's evidence for a regular increase in the velocity dispersion of the thin disk with decreasing galactocentric distance; in the central parts, this disk is hot enough to withstand bar instabilities. Mike Rich presented a very interesting study of K giants in the galactic bulge, which exhibit Fe/H ranging from -1 to almost +1; a simple model of galactic evolution can account for this observation.

Marc Aaronson reviewed the evidence for a strong intermediate-age population in the dwarf spheroidal and irregular galaxies. Clearly these systems--which may have some kind of relation with each other--have had a very different evolution from e.g. our Galaxy. Interesting and unexpected results have come from total-mass determinations in dwarf spheroidals and dwarf irregulars, obtained from the velocity dispersion of their stars or gas respectively: both types have mass/luminosity ratios ranging between 1 and 100: this was reported by Marc Aaronson and by Wal Sargent respectively. Thus, although many of these systems must have massive haloes (this is also evidenced by the ever-increasing rotation curve of some irregulars) there is a large spread in the properties of these haloes.

A large part of the meeting was devoted to star formation and to the initial mass function of stars in galaxies. Richard Larson reviewed the observations of the IMF and showed that its slope for massive stars is rather universal: $dn(m)/d \ln m \propto m^{-1.7\pm0.5}$, a fact also emphasized by Wendy Freedman (but the upper mass limit varies very much in different star-forming regions). This contrasts with the big variations in the luminosity function and hence in the IMF for low-mass stars in globular clusters as reported by Bob Lupton. Theory has made recent progresses (in particular some present ideas on the structure of the molecular clouds differ much from what they were only a few years ago) but we are still a long way of understanding the origin of the IMF. Larson also recalled his recent model of bimodal star formation in the Galaxy which has many attractive features although there may be problems with nucleosynthesis and chemical evolution in this model. Bob Kennicutt gave a comprehensive and well-balanced review of the determination of the star-formation rate and history in the disks of galaxies. While it is clear that this rate can be determined for high-mass stars at the present epoch (Angela Diaz showed in particular the interest of Balmer-line profile fitting in this respect) all the rest is still rather uncertain. A surprising fact is that the rate of massive star formation per unit mass is rather uniform in the disk of galaxies, at variance with the expectations of Schmidt's law. Reality must be different from a simple proportionality of the star formation rate to a power of the mass of gas: infall, or self-regulated star formation, must be invoked. The dwarf galaxies, reviewed by Wal Sargent, are even less well understood. We still do not know whether or not some of the blue compact galaxies are young (i.e. are presently experiencing their first burst of star formation), and we do not understand why their 21-cm structure can differ so much from one object to another, why the amount of dark matter is so variable, etc. Rather than saying again that Wal has already said better, I recommend reading Star-Forming Dwarf Galaxies, ed. D. Kunth, T. X. Thuan, J. T. T. Van, Editions Frontières, Gif-sur-Yvette (France), 1986.

Jim Truran was able to convey in a most clear way an enormous amount of information on nucleosynthesis; I retained from his talk that progress in this field is slow and that we are still plagued by uncertainties on major issues, e.g. the $^{12}C(\alpha, \gamma)^{16}O$ reaction rate. Most interesting was his discussion of abundances in Population II stars; the idea that the Population III required to produce initial enrichment in the extreme Population II could just be a first generation of high-mass stars in globular clusters may become more and more popular (there is a preprint by Roger Cayrel on that subject).

Most of the last day of this meeting was devoted to population synthesis in galaxies and its applications. Bob O'Connell described the techniques of spectral synthesis and their difficulties. In early-type galaxies, he shows that it should be possible to obtain the age of the last burst of star formation, essentially through an indirect determination of the turn-off point: but unfortunately this age determination depends on metallicity. More recent (< a few Gyr), lower-level star formation might be monitored by the UV excess as seen in many elliptical galaxies. However the nature of the stars responsible for this excess is not too clear. There is a feeling and even some direct indication for the nucleus of M31 from a recent rocket flight as reported by Bob that we are rather dealing with post-AGB stars. I am afraid that we will have to wait for the HST to be completely sure. Interpretation will require a better understanding of late stellar evolution; some surprises may come: Demarque reported in the discussion that new evolutionary models for very high metallicities predict an increased number of UV-bright stars. A nice place to check these ideas is the galactic bulge, although UV-bright stars are difficult to detect there due to heavy extinction; Albert Whitford showed however that there is no evidence in the bulge for any stars younger than a few, and perhaps 10 - 15 Gyr. Augustus Oemler reviewed the status of evolutionary studies of galaxies at high z. The Butcher-Oemler effect (bluer early-type galaxies in distant clusters) is now indirectly confirmed since galaxies in these distant clusters show more often emission lines or Balmer absorption-lines, revealing a hot population. The situation with radiogalaxies is controversial, a sad fact because they are easily found at large redshifts and might be better probes of the early times. Alvio Renzini then started to undermine some of the foundations of population synthesis, and a lively contest followed his talk. Whatever the issue, everyone agreed that we need to calibrate empirically the evolutionary tracks and in general the basic stellar data that are the inputs of synthesis models. The galactic bulge is almost the only place where we have at hand super-metallic stars. Unfortunately i) they are all old and ii) the large spread in metallicities is a severe difficulty. For solar and sub-solar metallicities the situation looks more favorable and much is to come from studies of the HR diagrams of clusters in the Galaxy, the LMC and the SMC. However the interplay in stellar evolution of age and of the various important chemical elements (He, C, N, O, Fe) is rather complex and it is very desirable that an independent direct determination of both these abundances and (if possible at all) of age is obtained before using these clusters as templates. This may be more difficult than observing their HR diagrams!

The general trend in this meeting is that galaxies are so difficult to understand that a prerequisite is a better understanding of our Galaxy and to nearest neighbors. This puts a renewed emphasis on direct stellar population studies. Fortunately CCD's came at the right time and a rapidly increasing amount of excellent CCD data is coming available: what was displayed in the posters at this workshop is only the tip of the iceberg. Spectral and radial-velocity surveys are also becoming increasingly numerous, a very welcome situation. Already the new data we have on cluster HR diagrams begin to form a good empirical basis for checking stellar evolution theories: indeed a good fit is starting to be possible for intermediate-mass and massive stars with mass-loss evolutionary models including convective overshooting: see e.g. the recent papers by Mermilliod and Maeder (Astron. Astrophys., **158**, 45, 1986), and Bertelli, Bressan and Chiosi (Astron. Astrophys., **150**, 33, 1985). Clearly study of stellar populations is one of the domains where the Hubble Space Telescope will have the largest impact. One easily foresees the following progresses:

i) deeper studies of clusters HR diagrams will be possible, including white dwarfs, blue HB and post-AGB stars etc. This will allow a deeper understanding of stellar evolution and provide better templates for population synthesis;

ii) the luminosity function and initial mass function will be obtained much more securely in clusters and in the field of nearby galaxies as a function of metallicity and other parameters; this will allow in particular some checking of the bimodal star-formation models, in particular the determination of lower mass cut-offs in the IMFs, if any;

iii) improved determinations of the abundances of heavy elements (CNO mainly) will be possible for hot cluster stars through high-resolution UV spectroscopy. This is much needed: for example, it is distressing to see that in the SMC (with only one unpublished exception to my knowledge, a K supergiant in NGC 330 studied by Spite and collaborators) the abundance determinations in SMC supergiants differ considerably, from each other and from the abundances in the gas. This is something very hard to understand for me at least. What is wrong? Gas mixing or abundances? This is one of the most interesting problems that can be solved with the HST.

DISCUSSION

WALBORN: I would like to stress how uncertain the upper region of the luminosity function (or mass function) is. The most massive stars tend to come in groups: for example in Carina there are close groups of several massive stars, in NGC 3603 there are about 12 stars within a few arc seconds, and the famous object R136 in the 30 Dor nebula contains no fewer than 27 stars in a radius of 4".5. Moreover, many new early O+WR stars have recently been found in the surroundings. HD 32228 is multiple, and there are many other examples in the LMC+SMC. These stars would be very difficult to separate in more distant galaxies, even with the HST. Hence the observed functions must be underestimating the numbers and overestimating the luminosoties of the most massive stars.

FREEDMAN: I would like to add that the surveys of luminous stars in nearby galaxies are anything but complete. It is not surprising that R136 has been shown to be composed of more than 1 star, since 1" corresponds to 0.2 pc at the distance of the LMC. But 1" is 3 pc at the distance of M33! The uppermost bin in the luminosity functions of the nearby galaxies shows a tremendous scatter relative to the fainter bins. This may be due in part to the small number of corresponding stars, but also may be a result of the multiplicity problem.

   I should point out that in the surveys the brightest stars in HII regions such as 30 Dor will not be picked up. I would stress that I think we are a long way from determining an absolute mass function in galaxies, or a correct upper mass limit. However I think that the fact that the colour-magnitude diagrams and the slopes of the luminosity functions for stars in nearby galaxies at different distances and with different metallicities and morphologies show such similarities in a relative sense is significant and is not likely due to some contrived sequence of compensating errors.

KING: The most interesting new thing that I have heard at this meeting was Truran's introduction of nucleosynthesis tracers into the development of globular clusters. We have been groping around for years in trying to understand their origin and their relative ages, and now there is a completely new line of potential evidence. We clearly need new work tracking the abundances of the key elements through the range of Fe/H, in globular-cluster stars and also in more halo subdwarfs.

EDMUNDS: The problem of the star and HII region abundances being different is not particular to the SMC. In our own Galaxy the Hyades are perhaps 0.2 dex up on the Sun and the Orion gas 0.3 dex down. It is unlikely that this is due to grains locking up elements in Orion since volatiles like neon are not overabundant. There just seems to be some fundamental calibration offset between analyses of young stars and HII gas. The difference between SMC young stars and Galaxy young stars is about the same as between the SMC HII regions and Orion. But the origin of the offset is unknown.

LEQUEUX: I agree as to the existence of this offset. However I do not quite agree on your last sentence about the SMC. For example while O/H is down by -0.7 with respect to Orion some young SMC stars are claimed to have solar oxygen abundances.

WALBORN: I would suspect observational or analytical incertainties in the various quantitative studies of SMC stars, because low-resolution spectrograms of early-type stars in the SMC show systematically and substantially weaker metal lines than those in the Galaxy or LMC.